Six Sigma Green Belt Exam Prep

500 Practice Questions

1st Edition

www.versatileread.com

Copyright © 2024 VERSAtile Reads. All rights reserved.
This material is protected by copyright, any infringement will be dealt with legal and punitive action.

Document Control

Proposal Name	:	Six Sigma Green Belt Exam Prep: 500 Practice Questions
Document Edition	:	1st
Document Release Date	:	9th July 2024
Reference	:	CLSSGB-001
VR Product Code	:	202424026SigmaGB

Copyright © 2024 VERSAtile Reads.
Registered in England and Wales
www.versatileread.com

All rights reserved. No part of this book may be reproduced or transmitted in any form or by any means, electronic or mechanical, including photocopying, recording, or by any information storage and retrieval system, without the written permission from VERSAtile Reads, except for the inclusion of brief quotations in a review.

Feedback:

If you have any comments regarding the quality of this book or otherwise alter it to better suit your needs, you can contact us through email at info@versatileread.com

Please make sure to include the book's title and ISBN in your message.

About the Contributors:

Nouman Ahmed Khan

AWS/Azure/GCP-Architect, CCDE, CCIEx5 (R&S, SP, Security, DC, Wireless), CISSP, CISA, CISM, CRISC, ISO27K-LA is a Solution Architect working with a global telecommunication provider. He works with enterprises, mega-projects, and service providers to help them select the best-fit technology solutions. He also works as a consultant to understand customer business processes and helps select an appropriate technology strategy to support business goals. He has more than eighteen years of experience working with global clients. One of his notable experiences was his tenure with a large managed security services provider, where he was responsible for managing the complete MSSP product portfolio. With his extensive knowledge and expertise in various areas of technology, including cloud computing, network infrastructure, security, and risk management, Nouman has become a trusted advisor for his clients.

Abubakar Saeed

Abubakar Saeed is a trailblazer in the realm of technology and innovation. With a rich professional journey spanning over twenty-nine years, Abubakar has seamlessly blended his expertise in engineering with his passion for transformative leadership. Starting humbly at the grassroots level, he has significantly contributed to pioneering the Internet in Pakistan and beyond. Abubakar's multifaceted experience encompasses managing, consulting, designing, and implementing projects, showcasing his versatility as a leader.

His exceptional skills shine in leading businesses, where he champions innovation and transformation. Abubakar stands as a testament to the power of visionary leadership, heading operations, solutions design, and integration. His emphasis on adhering to project timelines and exceeding customer expectations has set him apart as a great leader. With an unwavering commitment to adopting technology for operational simplicity and enhanced efficiency, Abubakar Saeed continues to inspire and drive change in the industry.

Dr. Fahad Abdali

Dr. Fahad Abdali is an esteemed leader with an outstanding twenty-year track record in managing diverse businesses. With a stellar educational background, including a bachelor's degree from the prestigious NED University of Engineers & Technology and a Ph.D. from the University of Karachi, Dr. Abdali epitomizes academic excellence and continuous professional growth.

Dr. Abdali's leadership journey is marked by his unwavering commitment to innovation and his astute understanding of industry dynamics. His ability to navigate intricate challenges has driven growth and nurtured organizational triumph. Driven by a passion for excellence, he stands as a beacon of inspiration within the business realm. With his remarkable leadership skills, Dr. Fahad Abdali continues to steer businesses toward unprecedented success, making him a true embodiment of a great leader.

Muniza Kamran

Muniza Kamran is a technical content developer in a professional field. She crafts clear and informative content that simplifies complex technical concepts for diverse audiences, with a passion for technology. Her expertise lies in Microsoft, cybersecurity, cloud security and emerging technologies, making her a valuable asset in the tech industry. Her dedication to quality and accuracy ensures that her writing empowers readers with valuable insights and knowledge. She has done certification in SQL database, database design, cloud solution architecture, and NDG Linux unhatched from CISCO.

Table of Contents

Overview of Six Sigma Green Belt 6
    What is Six Sigma? 6
    What is a Six Sigma Green Belt? 7
        Certified Lean Six Sigma Green Belt 7
        Certification Examination 8
        Requirements 8
        Preparation 8
        Certification 9
    Why Become a Certified Lean Six Sigma Green Belt? 9
    Benefits of Green Belt Certification 10
    Career Opportunities for a Certified Lean Six Sigma Green Belt 11
    What is the Salary of a Six Sigma Green Belt? 12
    Ace Six Sigma Certification 12
Practice Questions 13
Answers 118
About Our Products 203

Overview of Six Sigma Green Belt

In today's competitive business landscape, ensuring top-notch quality is paramount. **Six Sigma Green Belt Exam Cram: Essentials for Exam Success by VERSAtile Reads Belt** equips you with the key to unlocking this potential. Six Sigma, a powerful Process Improvement and Maintenance System, serves as your sixth sense for quality management. This guide unveils the path to Six Sigma Green Belt certification, empowering you to spearhead quality control initiatives within your organization. But before we delve into the specifics of Green Belt certification, let's break down the fundamental ABCs of Six Sigma, ensuring a smooth and successful journey towards quality excellence.

What is Six Sigma?

Six Sigma is a collection of methodologies and tools aimed at improving business processes by reducing defects and errors, minimizing variation, and enhancing quality and efficiency. The objective of Six Sigma is to reach a nearly perfect quality level, with only 3.4 defects per million opportunities. This is accomplished through a structured approach known as DMAIC (Define, Measure, Analyze, Improve, Control), which identifies and eliminates sources of variation to improve processes.

Six Sigma is a disciplined, data-driven methodology widely used in project management to improve processes and reduce defects. It provides a systematic framework for identifying and eliminating variations that can negatively impact project performance.

The term "Six Sigma" derives from the Greek symbol "sigma" (σ), which represents a statistical measure of process deviation from the mean or target. "Six Sigma" refers to the statistical concept illustrated by the bell curve, where one Sigma indicates a single standard deviation from the mean. When a process has six Sigmas—three above and three below the mean—the defect rate is considered "extremely low."

What is a Six Sigma Green Belt?

Lean Six Sigma Green Belts are professionals trained in Six Sigma Green Belt-level improvement methods who lead continuous process improvement groups as part of their regular responsibilities. This enables company leaders and champions to focus more on strategy-building and decision-making aspects of Lean Six Sigma project planning. Green Belts gain a comprehensive understanding of processes and collaborate closely with Subject Matter Experts (SMEs) to achieve performance goals.

Depending on the organization, Lean Six Sigma Green Belts may spend between 10-25% of their time on Lean Six Sigma projects. They are responsible for data collection and analysis within their project teams. Their facilitation skills and ability to lead root cause analysis sessions enhance team dynamics and effectiveness.

Lean Six Sigma Process Improvement Projects are applicable to every job function in every industry. Green Belt Certification is relevant across all sectors, whether manufacturing, sales, finance, marketing, or administrative roles. It broadens your process improvement toolkit and equips your team to select and apply the most appropriate tools for various situations.

Certified Lean Six Sigma Green Belt

The IASSC Certified Lean Six Sigma Green Belt™ (ICGB™) is a professional proficient in the core to advanced elements of Lean Six Sigma methodology. They lead improvement projects or serve as team members on more complex projects led by a Certified Black Belt, typically in a part-time capacity. A Lean Six Sigma Green Belt has a comprehensive understanding of all aspects of Lean Six Sigma, including the phases of Define, Measure, Analyze, Improve, and Control (DMAIC) as outlined by the IASSC Lean Six Sigma Green Belt Body of Knowledge™. They are skilled in implementing, performing, interpreting, and applying Lean Six Sigma at a high level of proficiency.

Overview of Six Sigma Green Belt

Certification Examination

The IASSC Certified Lean Six Sigma Green Belt Exam™ consists of 100 questions and is a closed-book, proctored examination with a duration of 3 hours. Certain versions of this exam may also include up to an additional 10 non-graded questions*. The exam comprises approximately 20 multiple-choice and true/false questions from each major section of the IASSC Lean Six Sigma Green Belt Body of Knowledge. It is available in over 8,000 Testing Centers across 165 countries worldwide, as well as through the PeopleCert Online Proctoring system and numerous IASSC Accredited Providers.

Requirements

To attain the professional designation of IASSC Certified Green Belt (IASSC-CGB™) from the International Association for Six Sigma Certification, candidates must take the IASSC Certified Lean Six Sigma Green Belt Exam and achieve a minimum score of 70%.

There are no prerequisites for sitting the IASSC Certified Lean Six Sigma Green Belt Exam. For further details, refer to the FAQ section and the IASSC guidelines on knowledge and application. Purchase a Green Belt Exam Voucher for USD 350 through the provided link.

Preparation

While it is recommended, it is not mandatory to undergo Lean Six Sigma training from a reputable institution, trainer, or corporate program before taking the exam. Similarly, it is advisable, though not obligatory, for candidates to possess some level of real-world Lean Six Sigma experience and project application background. IASSC provides a non-proctored (informal) Green Belt Evaluation Exam, which can help individuals assess their readiness for the official proctored Certification Exam.

Overview of Six Sigma Green Belt

Certification

Upon successful completion, Professionals will receive IASSC Green Belt Certification issued by the International Association for Six Sigma Certification™, an independent third-party Certification Association in the Lean Six Sigma Industry.

IASSC Certification remains valid indefinitely. As per the IASSC Recertification Policy implemented on March 1, 2017, certifications are classified as "Current" for three years, with guidelines for maintaining this status.

An IASSC Certified Green Belt receives a Certificate (PDF) containing a Certification Number and a badge (.png file), is added to the Official IASSC Certification Register, and is granted permission to use the IASSC Certification Marks and Titles. This professional designation can be included in resumes and public profiles such as LinkedIn, in accordance with the IASSC Marks Use Policy.

Why Become a Certified Lean Six Sigma Green Belt?

If you are interested in enhancing processes and minimizing errors in your organization, pursuing a Lean Six Sigma Green Belt certification is a wise choice. Whether you're new to Lean Six Sigma or aiming to advance your certified career, this guide provides all the essential information you need. Lean Six Sigma Green Belts (SSGBs) are professionals with a thorough understanding of advanced Six Sigma methodologies. They lead improvement projects and also serve as team members on more complex projects managed by Black Belts and Master Black Belts.

Green Belts are highly skilled team members dedicated to improving process quality. They bridge the gap between Six Sigma theory and its practical application, playing a crucial role in project management, process improvement, and data analysis. Green Belt certification training covers the fundamentals of project management and the application of DMAIC skills in Six Sigma projects.

Lean Six Sigma Green Belt certification is ideal for those with experience in supply chain management and business management, particularly those interested in concepts like waste reduction and continuous improvement. This certification enhances a candidate's proficiency, making them more attractive to employers. It is intended for professionals responsible for improving output, controlling costs, and enhancing results.

Benefits of Green Belt Certification

- **Professional Development**: Enhances problem-solving skills and the ability to make effective recommendations, boosting confidence in managing complex issues.
- **Financial Advantages:** Lowers project costs, improves project efficiency, and eliminates the need for reserve funds.
- **Customer Benefits:** Enhances product and service quality, fostering customer loyalty and attracting new clients who seek Six Sigma standards.
- **Competitive Edge:** Boosts business performance, supports marketing efforts, and maintains a competitive advantage by effectively implementing Six Sigma principles.

Career Opportunities for a Certified Lean Six Sigma Green Belt

The job duties of Green Belts can vary depending on the industry, but they generally focus on project management, process improvement, and data analysis. Professionals with Six Sigma Green Belt certification can pursue various roles, including:

- **Project Manager:** Responsible for prioritizing and leading projects of different sizes and complexities.
- **Six Sigma Consultant:** Works closely with the operations team to analyze and improve current processes using Project Management skills.
- **Reliability Engineer:** Develops reliability test plans and collaborates with customers and internal teams to resolve reliability issues.
- **Process Engineer:** Supports company leadership in evaluating internal processes for efficiency and quality.
- **Mechanical Engineer:** Develops new tools, devices, and products, using Six Sigma knowledge to create concepts, design equipment, and test products.
- **Production Engineer:** Optimizes and develops processes for designing, manufacturing, and shipping products.
- **Industrial Engineer:** Evaluates production rates, quality control processes, and employee productivity to enhance resource efficiency.
- **Continuous Improvement Specialist:** Identifies, coordinates, and conducts root cause analyses by collecting data.
- **Quality Engineer:** Ensures that a company's products meet customer requirements.
- **Operational Excellence Manager:** Manages Lean Six Sigma initiatives within an organization, working with team members at all levels, and reports directly to senior management.

What is the Salary of a Six Sigma Green Belt?

Companies worldwide seek Lean Six Sigma Green Belts to reduce costs and eliminate waste. Green Belts are considered valuable investments by employers because they are trained to use Lean Six Sigma tools and methodologies to prevent defects and enhance customer satisfaction.

Salaries can vary depending on the specific Green Belt role. For instance, a Process Engineer earns an average of $89,953 annually, a Quality Manager makes $93,199, a Continuous Improvement Manager earns $133,790, and a Quality Assurance or Quality Control Engineer makes $105,090 per year.

Ace Six Sigma Certification

Looking to advance your career and enhance your skills? Versatile Read's Exam Cram: Six Sigma Green Belt offers comprehensive knowledge and tools to help you excel in process improvement within your organization.

Leveraging the proven DMAIC methodology, you'll gain the expertise to lead and execute impactful projects that streamline operations, minimize defects, and, ultimately, enhance customer satisfaction. Not only will you master the fundamentals of Lean Six Sigma, but you'll also delve into effective team management strategies and delve into strategic tools for meticulous performance measurement and analysis.

Aligned with the exam blueprints, this comprehensive guide prepares you to achieve global recognition as a certified Lean Six Sigma Green Belt professional. This certification opens doors to exciting career opportunities and positions you as a valuable asset in today's quality-driven business landscape.

Practice Questions

Practice Questions

1. Why must every organization, including not-for-profits, have a source of income?
    A. To pay taxes
    B. To stay in business
    C. To expand internationally
    D. To attract investors

2. What is the first formula that every Six Sigma Green Belt must learn?
    A. Return on Investment (ROI).
    B. Net Present Value (NPV).
    C. S-double bar.
    D. Internal Rate of Return (IRR).

3. Which of the following is NOT a phase in the DMAIC methodology?

    A. Define

    B. Measure

    C. Act

    D. Improve

4. What is a common reason for the failure of Six Sigma implementation in organizations?
    A. Lack of external consultants
    B. Excessive training of Yellow Belts
    C. Management's lack of commitment to real process improvement
    D. Non-involvement of customers

5. What typically happens when consultants leave after initial Six Sigma training?
    A. Projects continue to thrive

Practice Questions

   B. Internal people understand and maintain the tools effectively
   C Few internal people understand the tools at high enough levels
   D. The organization hires new consultants immediately

6. What key department must be engaged early in Six Sigma deployment to substantiate cost savings?
   A. Human Resources
   B. Marketing
   C. Accounting
   D. Research and Development

7. Which video from the mid-1950s is recommended to understand the importance of getting things right first?
   A. Quality Control in Manufacturing
   B. The Process of Six Sigma
   C. Right First Time
   D. Efficient Production Techniques

8. What title do some organizations give to a highly competent Six Sigma professional?
   A. Six Sigma Expert
   B. Six Sigma Money Belt
   C. Six Sigma Champion
   D. Six Sigma Executive

9. What should management use to prepare for a Six Sigma deployment to increase the likelihood of success?
   A. Advanced Quality Planning (AQP)
   B. Total Quality Management (TQM)
   C. Lean Manufacturing
   D. Statistical Process Control (SPC)

Practice Questions

10. Who was engaged to improve performance in doctors' offices?
    A. Nurses
    B. Quality engineers and process engineers
    C. Medical students
    D. Hospital administrators

11. What is the primary function of the Taguchi loss function?
    A. Measure financial loss to society resulting from poor quality
    B. Identify critical inputs in a process
    C. Control charting
    D. Define DMAIC stages

12. Which of the following was the first quality control chart used?
    A. p-chart
    B. c-chart
    C. u-chart
    D. np-chart

13. What is the main goal of the Six Sigma philosophy?
    A. Increasing product variety
    B. Reducing process variation
    C. Enhancing marketing strategies
    D. Expanding market presence

14. Which quality improvement approach involves restructuring an entire organization and its processes?
    A. Benchmarking
    B. Reengineering
    C. Six Sigma
    D. ISO 9000

Practice Questions

15. What does the acronym DMAIC stand for in Six Sigma methodology?
    A. Design, Measure, Analyze, Improve, Control
    B. Define, Measure, Analyze, Improve, Control
    C. Develop, Measure, Apply, Implement, Control
    D. Define, Manage, Assess, Implement, Control

16. What concept is associated with contributions to quality management?
    A. Quality circles
    B. Red "X"
    C. Juran trilogy
    D. Taguchi loss function

17. Which international standard focuses on quality management and assurance?
    A. ISO 14000
    B. ISO 45001
    C. ISO 9000
    D. ISO 31000

18. What is the key focus of Lean manufacturing, often associated with Six Sigma?
    A. Market Expansion
    B. Waste reduction
    C. Product diversification
    D. Enhancing marketing strategies

19. Which of the following is NOT one of Feigenbaum's three steps to quality?
    A. Quality leadership
    B. Modern quality technology
    C. Organizational commitment
    D. Financial investment

Practice Questions

20. Which award was established by the U.S. Congress in 1987 to recognize companies with successful quality management systems?
    A. ISO 9001
    B. Malcolm Baldrige National Quality Award
    C. Deming Prize
    D. Six Sigma Award

21. What is the primary focus of lean thinking?
    A. Increasing profitability
    B. Reducing employee workload
    C. Making obvious what adds value by reducing everything else
    D. Enhancing marketing efforts

22. Which company is usually credited with the creation of the concept of lean?
    A. Ford Motor Company
    B. General Motors
    C. Toyota
    D. Honda

23. Which of the following is NOT one of the 8Ps of lean thinking?
    A. Pull
    B. Perfection
    C. Prevention
    D. Profit

24. Which term is used to describe non-value-added activities in lean thinking?
    A. Muda
    B. Kaizen
    C. Jidoka
    D. Heijunka

Practice Questions

25. What are the three criteria for a process step to be considered value-added in lean?
    A. It is done quickly, it is efficient, it is low-cost
    B. The customer recognizes the value, it changes the product, it is done right the first time
    C. It uses advanced technology, it is simple, it has a high-profit margin
    D. It is standardized, it is documented, it is automated

26. Which principle of The Toyota Way emphasizes the importance of making decisions slowly by consensus?
    A. Go and see for yourself
    B. Level out the workload
    C. Make decisions slowly by consensus
    D. Become a learning organization

27. What is the primary goal of the 5S methodology in lean?
    A. Increase sales
    B. Reduce costs
    C. Improve workplace organization and efficiency
    D. Enhance customer satisfaction

28. Which lean principle involves stopping the production process to fix problems to ensure quality the first time?
    A. Jidoka
    B. Heijunka
    C. Kaizen
    D. Poka-yoke

29. Which of the following is NOT a component of the TPS house?
    A. Just-in-time
    B. Jidoka
    C. Kaizen
    D. Six Sigma

Practice Questions

30. In the context of lean, what does the term "takt time" refer to?
    A. The time taken for a machine to complete a cycle
    B. The rate at which products need to be produced to meet customer demand
    C. The time employees spend on breaks
    D. The overall production time in a shift

31. Which of the following is the most difficult sequence in the 5S methodology?
    A. Sort
    B. Set in Order
    C. Shine
    D. Sustain

32. What is the primary purpose of the Andon system in a lean manufacturing environment?
    A. To reduce inventory levels
    B. To alert operators and supervisors of the production status
    C. To standardize work procedures
    D. To organize the workplace

33. What is the main goal of Hoshin Kanri?
    A. To create a visual management system
    B. To align company goals with middle management tactics and operational actions
    C. To reduce lead times and inventory levels
    D. To automate repetitive tasks

34. In the context of 5S, what does the 'Shine' step involve?
    A. Eliminating unnecessary items
    B. Organizing remaining items
    C. Cleaning and inspecting the work area

Practice Questions

   D. Standardizing cleaning practices

35. What is a Kaizen event typically characterized by?
    A. Long-term, gradual improvements
    B. Quick, breakthrough improvements
    C. Reducing inventory levels
    D. Standardizing work procedures

36. What does the term 'Jidoka' refer to in lean manufacturing?
    A. Continuous improvement
    B. Just-in-time production
    C. Intelligent automation with a human touch
    D. Visual management

37. Which of the following is a key element in ensuring the success of the 5S methodology according to John Casey?
    A. Starting with the 'Sort' step
    B. Engaging management in cost savings
    C. Focusing on inventory reduction
    D. Implementing a pull system

38. What is the purpose of a Kanban system in lean manufacturing?
    A. To create a continuous flow
    B. To implement policy deployment
    C. To act as a pull system for inventory control
    D. To perform quick status reports

39. Which principle promotes understanding real-world operational issues by firsthand observation?
    A. Kanban
    B. Hoshin Kanri
    C. Gemba

Practice Questions

D. Heijunka

40. What additional 'S' is sometimes added in healthcare to the 5S methodology?
    A. Safety
    B. Standardize
    C. Shine
    D. Oversight

41. What is the primary purpose of a Kanban system?
    A. To measure the overall effectiveness of the operation
    B. To prevent errors in the production process
    C. To ensure timely movement of products and information
    D. To reduce production lot sizes

42. Which of the following is NOT a component of Standard Work?
    A. Standard time
    B. Standard inventory
    C. Standard sequence
    D. Standard cost

43. What does the term 'takt time' refer to in a lean production system?
    A. The maximum time allowed for a single process step
    B. The time it takes to change over to a different product
    C. The pace at which production should occur to meet customer demand
    D. The total time from raw material to finished product

44. What is the focus of the Theory of Constraints?
    A. Eliminating all forms of waste in production
    B. Reducing setup times to less than 10 minutes
    C. Improving the rate of the slowest process in a chain

Practice Questions

    D. Standardizing work processes

45. In the Drum-Buffer-Rope analogy, what does the 'drum' represent?
    A. The fastest process in the system
    B. The slowest process that sets the pace for the entire system
    C. The total amount of work-in-process (WIP)
    D. The total lead time required to complete production

46. What does Poka-Yoke aim to achieve in a production process?
    A. Reduce cycle times
    B. Prevent errors and mistakes
    C. Increase takt time
    D. Maximize equipment utilization

47. What does Overall Equipment Effectiveness (OEE) measure?
    A. Economic lot size
    B. Total production time
    C. Effectiveness of a manufacturing operation
    D. Changeover time

48. What is the goal of Single Minute Exchange of Die (SMED)?
    A. To standardize work processes
    B. To prevent production errors
    C. To reduce changeover times to less than 10 minutes
    D. To measure equipment effectiveness

49. Which lean concept emphasizes the importance of visual indicators to manage the status of material and information?
    A. Visual Factory
    B. Poka-Yoke
    C. Takt Time
    D. Theory of Constraints

Practice Questions

50. What role does Total Productive Maintenance (TPM) play in a lean enterprise?
    A. It standardizes work processes
    B. It reduces variation in the process
    C. It ensures equipment reliability and reduces downtime
    D. It measures the effectiveness of the operation

51. What is the principal symptom of overproduction in a manufacturing process?
    A. Long setup times
    B. Excess work-in-process (WIP)
    C. Poor maintenance
    D. Delayed shipments

52. Which of the following is a cause of excess motion waste?
    A. Delayed shipments
    B. Poor workplace layout
    C. Long setup times
    D. High defect rates

53. What is a common cause of waiting waste in a production environment?
    A. Poor training
    B. Delayed shipments
    C. Overproduction
    D. Excessive material movement

54. Why is maintaining high levels of inventory considered wasteful?
    A. It requires more cross-training
    B. It leads to excessive motion
    C. It adds costs without adding value
    D. It improves quality levels

Practice Questions

55. What is the primary cause of the excess movement of material/transportation waste?
    A. Poor plant layout
    B. Long setup times
    C. High inventory levels
    D. Delayed shipments

56. What is the effect of defect correction in the production process?
    A. It adds value to the product
    B. It wastes effort
    C. It reduces waiting time
    D. It improves ergonomic conditions

57. What problem does a "sea of inventory" analogy illustrate?
    A. It hides unresolved problems
    B. It reduces waiting time
    C. It improves product quality
    D. It enhances employee training

58. Which strategy is recommended to reduce the waste of waiting?
    A. Increase inventory
    B. Implement Kaizen events
    C. Cross-train personnel
    D. Increase overproduction

59. Which of the following is a symptom of an unbalanced workload?
    A. Delayed shipments
    B. Excess WIP
    C. High defect rates
    D. Poor equipment maintenance

Practice Questions

60. What is an example of excess processing waste?
    A. Painting parts immediately after stamping
    B. Dipping parts in oil to prevent rust before painting
    C. Cross-training employees for multiple tasks
    D. Implementing a just-in-time production system

61. What is the primary goal of eliminating 'muda' in lean thinking?
    A. To increase production speed
    B. To strive for perfection
    C. To reduce employee workload
    D. To enhance marketing strategies

62. Which step in the value stream mapping process involves analyzing all inventory notes with an eye toward reduction or elimination?
    A. Produce a value stream map
    B. Analyze all inventory notes
    C. Extend the value stream map upstream
    D. Determine how the flow is driven

63. What is the purpose of the grinding step that often follows a welding operation?
    A. To add a protective layer
    B. To remove weld imperfections
    C. To increase the material's hardness
    D. To prepare for painting

64. Which of the following is considered a form of waste in lean thinking?
    A. Efficient communication
    B. Lost creativity
    C. Increased quality control
    D. Optimized production process

Practice Questions

65. What does the acronym DMADV stand for in Design for Six Sigma?
    A. Define, Measure, Analyze, Design, Verify
    B. Define, Measure, Assess, Design, Validate
    C. Develop, Measure, Analyze, Design, Verify
    D. Design, Measure, Analyze, Develop, Verify

66. What is the purpose of the 'Identify' phase in the IDOV process?
    A. To formulate concept design
    B. To prototype test and validation
    C. To link the design to the voice of the customer
    D. To optimize design to minimize sensitivity

67. Which of the following is NOT a component of a value stream?
    A. Flow of materials
    B. Transformation of raw materials
    C. Flow of information
    D. Customer purchase decisions

68. What term is used to describe unnecessary grinding after a welding operation?
    A. Excess inventory
    B. Excessive processing
    C. Lost creativity
    D. Overproduction

69. Which phase in the IDOV process focuses on optimizing the design to minimize sensitivity of CTQs to process parameters?
    A. Identify
    B. Design
    C. Optimize
    D. Validate

Practice Questions

70. What is a value stream map used to illustrate?
    A. Marketing strategies
    B. Financial projections
    C. Movement of material, information, inventory, and work-in-process
    D. Employee performance metrics

71. What is the primary goal of Design for Cost (Design to Cost) in DFSS?
    A. To improve product performance
    B. To ensure products are easy to test
    C. To reduce manufacturing costs
    D. To enhance product maintainability

72. Which term refers to the design approach that ensures products can be easily tested during the production cycle?
    A. Design for Manufacturing
    B. Design for Testability
    C. Design for Maintainability
    D. Design for Scalability

73. In DFSS, what does MTTF stand for, and what is it used for?
    A. Mean Time to Failure; used for repairable items
    B. Mean Time to Failure; used for non-repairable items
    C. Mean Time to Fix; used for repairable items
    D. Mean Time to Fix; used for non-repairable items

74. What does the term "Design for Scalability" refer to in the context of DFSS?
    A. Ensuring the product can be easily maintained
    B. Allowing for the product to be expanded or rapidly adopted in a growth market
    C. Making the product secure from unauthorized use
    D. Designing the product to be cost-efficient

Practice Questions

75. Which DFSS principle focuses on the ease of performing routine maintenance on a product?
    A. Design for Usability
    B. Design for Agility
    C. Design for Maintainability
    D. Design for Performance

76. What is the main purpose of a Failure Mode and Effects Analysis (FMEA)?
    A. To improve product usability
    B. To identify potential failure causes and their effects
    C. To enhance product aesthetics
    D. To streamline product design

77. In FMEA, what does the term "Risk Priority Number (RPN)" represent?
    A. The sum of severity, occurrence, and detection scores
    B. The highest severity score
    C. The product of severity, occurrence, and detection scores
    D. The average of severity, occurrence, and detection scores

78. Which DFSS principle ensures that products are designed to consume minimal resources?
    A. Design for Usability
    B. Design for Efficiency
    C. Design for Performance
    D. Design for Security

79. What is a key benefit of involving manufacturing personnel early in the design process?
    A. Enhanced product aesthetics
    B. Improved product scalability
    C. Greater producibility and reduced manufacturing costs
    D. Increased product security

Practice Questions

80. Which DFSS principle focuses on the ability to meet aggressive benchmarks or breakthrough performance levels consistently?
    A. Design for Usability
    B. Design for Agility
    C. Design for Performance
    D. Design for Compliance

81. Which of the following is NOT typically included in a Six Sigma project proposal?
    A. Preliminary measures of the seriousness of the problem
    B. Exact financial projections
    C. Precise statements of the problem definition
    D. Impact on organizational goals

82. What is the primary factor for gauging both the performance and health of an organization in Six Sigma projects?
    A. Number of completed projects
    B. Customer satisfaction
    C. Use and selection of metrics
    D. Employee feedback

83. Which of the following tools can be used for project selection in an organization that does not have a selection process in place?
    A. SWOT analysis
    B. Advanced quality planning and quality function deployment
    C. PERT charts
    D. Business process reengineering

84. Which type of benchmarking involves comparing similar functions outside the organization's industry?
    A. Internal benchmarking
    B. Competitive benchmarking

Practice Questions

   C. Functional benchmarking
   D. Collaborative benchmarking

85. What does the SIPOC model stand for?
    A. Suppliers, Inputs, Processes, Outputs, Customers
    B. Systems, Inputs, Processes, Outputs, Control
    C. Suppliers, Inputs, Production, Outputs, Control
    D. Systems, Information, Processes, Outputs, Customers

86. What is the primary purpose of including stakeholders in process improvement teams?
    A. To increase the number of team members
    B. To comply with organizational policy
    C. To utilize their knowledge and ensure their buy-in
    D. To provide financial support

87. Which of the following is a key activity in quality function deployment (QFD)?
    A. Developing a marketing strategy
    B. Linking customer requirements to product features
    C. Conducting financial audits
    D. Training employees on safety protocols

88. What is the primary goal of benchmarking in Six Sigma projects?
    A. To document all company processes
    B. To achieve breakthrough improvement
    C. To maintain the status quo
    D. To reduce the workforce

89. What should be done after collecting customer data in a Six Sigma project?
    A. Store the data for future reference

Practice Questions

B. Verify the accuracy and consistency of the data
C. Present the data to the board of directors
D. Discard any conflicting data

90. When defining a process, why is it important to identify its start and end points?
    A. To ensure it fits into the company's budget
    B. To determine the number of employees needed
    C. To clearly define the process boundaries
    D. To align with the company's marketing strategy

91. What does a negative co-relationship in a QFD matrix indicate?
    A. Both technical requirements improve together.
    B. Improving one technical requirement makes the other worse.
    C. Both technical requirements worsen together.
    D. There is no relationship between the technical requirements

92. What is the purpose of the column weights in a QFD matrix?
    A. To indicate the cost of technical requirements.
    B. To show the frequency of customer complaints.
    C. To reflect the importance of technical requirements in meeting customer requirements.
    D. To measure the duration of each technical requirement

93. Which symbol represents a strong relationship in a QFD matrix?
    A. 1
    B. 3
    C. 5
    D. 9

94. What is the primary purpose of Quality Function Deployment (QFD)?
    A. To reduce manufacturing costs.
    B. To increase product sales.

Practice Questions

    C. To embed customer requirements into the earliest stages of product development.
    D. To standardize production processes.

95. What does the term "Whats" refer to in a QFD matrix?
    A. Technical requirements
    B. Customer requirements
    C. Production costs
    D. Project timelines

96. Which quality tool is recommended for condensing a list of more than 25 customer voice lines in a QFD matrix?
    A. Fishbone diagram
    B. Pareto chart
    C. Affinity diagram
    D. Control chart

97. What should a project charter include to help the team stay on track with organizational goals?
    A. A list of all company employees
    B. A detailed budget breakdown
    C. Problem statement, purpose, benefits, scope, and results
    D. Daily work schedules

98. Which project planning tool is helpful for visualizing the sequence and duration of project tasks?
    A. Fishbone diagram
    B. Gantt chart
    C. Control chart
    D. Histogram

99. What is a Risk Priority Number (RPN) used for in project risk analysis?
    A. To calculate project costs

Practice Questions

   B. To prioritize risks based on their severity and probability
   C. To determine customer satisfaction levels
   D. To schedule project tasks

100. In the context of QFD, what does the relationship matrix map?
   A. Customer requirements to technical requirements
   B. Project costs to project timelines
   C. Employee tasks to project milestones
   D. Supplier capabilities to production schedules

101. What is the first step in risk mitigation?
   A. Implementing mitigation plans
   B. Identifying mitigation activities
   C. Reviewing the status of risks
   D. Verifying risk mitigation activities

102. Which of the following is an example of a risk mitigation activity for a key supplier not delivering on time?
   A. Ignoring the risk
   B. Developing another supplier
   C. Reducing the project scope
   D. Increasing project funding

103. How often should risk mitigation plans be reviewed?
   A. Annually
   B. Bi-annually
   C. Weekly or monthly
   D. Only at the end of the project

104. What is the purpose of risk verification?
   A. To identify new risks
   B. To ensure the project stays on budget

Practice Questions

C. To ensure that risk mitigation activities prevent the risk from occurring
D. To close the project

105. What should be done if risk mitigation is not possible and the risk is realized?
    A. Ignore the risk
    B. Implement the pre-planned response
    C. Start a new project
    D. Cancel the project

106. Which of the following best describes the process of project closure?
    A. Negotiating final project terms
    B. Conducting a project audit
    C. Reviewing project objectives achieved in relation to the charter
    D. Continuing the project indefinitely

107. What is the role of the project charter in project closure?
    A. It is used to measure project completion
    B. It is discarded after the project starts
    C. It is used to request additional funding
    D. It is used to negotiate project terms

108. What is typically sufficient for closing a project?
    A. An independent review
    B. An audit approach
    C. A review of the charter against documented project results
    D. A project extension

109. What should be identified and informed to other parts of the organization during project closure?
    A. New project opportunities

Practice Questions

B. Lessons learned and opportunities for improvement
C. Budget surpluses
D. New team members

110. What might a project sponsor request prior to formal project closure?
    A. Additional funds
    B. An independent review or assessment
    C. A new project plan
    D. A new risk assessment

111. Which tool graphically shows interdependencies between tasks and can provide either a top-level overview or detailed data?
    A. Affinity Diagram
    B. Activity Network Diagram (AND)
    C. Check Sheets
    D. Customer Feedback

112. Which tool is designed to produce many possible answers to an open-ended question by organizing brainstormed responses into natural groupings?
    A. Advanced Quality Planning (AQP)
    B. Auditing
    C. Affinity Diagram
    D. Benchmarking

113. What is the primary purpose of Advanced Quality Planning (AQP)?
    A. To identify vendor communication issues
    B. To ensure that all elements/inputs are considered to achieve the desired output
    C. To provide detailed customer feedback
    D. To create check sheets for data collection

Practice Questions

114. What is a key step in the benchmarking process?
    A. Create a cause-and-effect diagram
    B. Pilot test ideas
    C. Perform a safety audit
    D. Formulate and publish RFP

115. Which of the following is NOT one of the six Ms used in creating a cause-and-effect diagram?
    A. Money
    B. Methods
    C. Management
    D. Marketing

116. What is the primary purpose of auditing within an organization?
    A. To collect customer feedback
    B. To independently assess processes or projects against established plans and goals
    C. To brainstorm potential improvements
    D. To create activity network diagrams

117. Which of the following is an essential aspect of conducting an effective audit?
    A. Being overly critical
    B. Hiding observed issues
    C. Being thorough in looking at the process as it relates to the standard
    D. Ignoring the internal lead auditor's directions

118. Which tool involves collecting data that can be used for further analysis and identifying common patterns or trends?
    A. Check Sheets
    B. Affinity Diagram

Practice Questions

C. Cause-and-Effect Diagram
D. Benchmarking

119. What is one potential downside of traditional customer feedback surveys?
   A. They are too detailed
   B. They have a very high return rate
   C. They often do not provide a very good overall picture
   D. They are too expensive to implement

120. What is the first step when starting a benchmarking project?
   A. Develop plans for the application of ideas
   B. Flowchart of the current process
   C. Brainstorm ideas
   D. Investigate how others perform similar processes

121. What is the first step in developing a flowchart, according to Nancy Tague in The Quality Toolbox?
   A. Discuss and decide on the boundaries of your process
   B. Define the process to be diagrammed
   C. Brainstorm the activities that take place
   D. Arrange the activities in proper sequence

122. Which method is considered most effective for understanding customer needs?
   A. Conducting surveys
   B. Analyzing sales data
   C. Getting out into the world with the customers
   D. Reviewing customer complaints

123. In a flowchart, what does an arrow typically represent?
   A. A decision point
   B. The start of the process

Practice Questions

    C. The flow of the process
    D. The end of the process

124. Which of the following is NOT a type of control chart mentioned?
    A. X- and R chart
    B. EWMA chart
    C. Paynter chart
    D. Benchmark chart

125. What does the term 'driver' refer to in an interrelationship digraph?
    A. The most influential factor
    B. The least influential factor
    C. The outcome
    D. The process start point

126. Which technique is best for gaining accurate, in-depth customer responses?
    A. Surveys
    B. Focus groups
    C. Interviews
    D. Observation

127. What does the 'Check' phase of the PDCA cycle involve?
    A. Implementing a solution
    B. Evaluating the results
    C. Defining a problem
    D. Standardizing the solution

128. In a prioritization matrix, what does the team do after ranking each option against the criteria?
    A. Assign weights to the criteria
    B. Calculate the row totals

Practice Questions

C. Discuss possible solutions
D. Draw arrows to show relationships

129. Which tool is described as being used to illustrate the steps of a process in sequential order?
    A. Cause-and-effect diagram
    B. Flowchart
    C. Control chart
    D. Gantt chart

130. What is the primary purpose of Failure Mode and Effects Analysis (FMEA)?
    A. To prioritize tasks
    B. To identify potential process failures and their effects
    C. To improve customer satisfaction
    D. To create control charts

131. Which of the following steps is the first in the eight-discipline approaches (8D) to problem-solving?
    A. Describe the problem
    B. Use a team approach
    C. Start and check interim (containment) actions
    D. Define and check root causes

132. What is the purpose of the Project Evaluation and Review Technique (PERT)?
    A. To prioritize the most suitable software package
    B. To illustrate anticipated problems and list possible solutions
    C. To identify and calculate the critical path for projects
    D. To define and implement containment actions

Practice Questions

133. In the context of sampling, what is the accuracy percentage of 100 percent inspection as stated by Dr. Joseph Juran?
    A. 60%
    B. 70%
    C. 80%
    D. 90%

134. What does the Risk Priority Number (RPN) represent in Failure Mode and Effects Analysis (FMEA)?
    A. The sum of severity, occurrence, and detection ratings
    B. The product of severity, occurrence, and detection ratings
    C. The average of severity, occurrence, and detection ratings
    D. The maximum of severity, occurrence, and detection ratings

135. What is the purpose of a tree diagram in problem solving?
    A. To identify the most suitable software package
    B. To illustrate anticipated problems and list possible solutions
    C. To break a general topic into a series of more specific topics
    D. To define and implement containment actions

136. Which of the following is a measure of process performance that calculates defects per unit (DPU)?
    A. The total number of defects divided by the total number of products produced
    B. The total number of opportunities for defects divided by the total number of products produced
    C. The total number of products produced divided by the total number of defects
    D. The total number of opportunities for defects divided by the total number of defects

137. How is the rolled throughput yield (RTY) calculated?
    A. By multiplying individual process yields

Practice Questions

   B. By adding individual process yields
   C. By dividing the total yield by the number of processes
   D. By subtracting the defect rate from 1

138. Which of the following communication flows is used when top management provides instructions or feedback on project performance?
   A. Bottom-up flow
   B. Horizontal communication
   C. Top-down flow
   D. Diagonal communication

139. Which quality management approach emphasizes full participatory management?
   A. Juran trilogy
   B. Total quality control
   C. DMAIC
   D. Respect for humanity

140. Which of the following is a key contribution of Genichi Taguchi?
   A. Ishikawa diagram
   B. Quality circles
   C. Off-line quality control
   D. Juran trilogy

141. What is considered an ideal team size for effective teamwork and interaction?
   A. 3 members
   B. 7 members
   C. 11 members
   D. 15 members

Practice Questions

142. Which stage of team development is characterized by members resolving their conflicts and starting to trust each other?
    A. Forming
    B. Storming
    C. Norming
    D. Performing

143. What is the main challenge associated with virtual teams?
    A. Reduced costs
    B. Real-time data sharing
    C. Slowing of team-building progression
    D. Increased face-to-face interaction

144. Which team tool involves generating a large number of creative ideas in a short period without criticism or judgment?
    A. Nominal Group Technique
    B. Multi-voting
    C. Brainstorming
    D. Affinity Diagram

145. In which stage of team development does the team create synergy and realize interdependence?
    A. Forming
    B. Storming
    C. Norming
    D. Performing

146. What is the primary responsibility of a Six Sigma Black Belt?
    A. Set overall business strategy
    B. Lead and manage Six Sigma projects
    C. Provide executive sponsorship
    D. Allocate resources for projects

Practice Questions

147. Which leadership style is appropriate during the Storming stage of team development?
    A. Directing
    B. Coaching
    C. Supporting
    D. Delegating

148. What is the main function of the Recognition stage in team dynamics?
    A. To resolve conflicts
    B. To disband the team
    C. To recognize and appreciate team achievements
    D. To set team goals

149. Which method is used to prioritize a list of ideas generated during brainstorming?
    A. Nominal Group Technique
    B. Multi-voting
    C. Affinity Diagram
    D. Fishbone Diagram

150. What is an effective countermeasure for dealing with dominant team members?
    A. Ignoring their input
    B. Conducting the meeting in a more informal setting
    C. Providing equal participation opportunities
    D. Letting them lead the discussion

151. What is the primary difference between process mapping and flowcharting?
    A. Process mapping uses only symbols while flowcharting uses text
    B. Process mapping includes additional process details with the flowchart

Practice Questions

C. Flowcharting is done before process mapping
D. Flowcharting includes costs, setup time, and types of defects

152. What is the first step in improving a process?
     A. Conducting a training session
     B. Implementing new software
     C. Process mapping
     D. Hiring a new manager

153. What international standard is mentioned in the context for documentation symbols and conventions for data, program, and system flowcharts?
     A. ISO 9000
     B. ISO 14001
     C. ISO 5807:1985
     D. ISO 31000

154. In a swim lane flowchart, how are the process blocks arranged?
     A. Randomly
     B. By time sequence
     C. In alignment with the lane of the department or function that performs a given process step
     D. By importance of the process step

155. What is a common mistake in process mapping as mentioned in the text?
     A. Using too many colors
     B. Not including enough decision points
     C. Team representation for the process is inadequate or inappropriate
     D. Over-documenting the process

Practice Questions

156. Which type of process map helps visualize process steps involving multiple departments or functions?
    A. Basic flowchart
    B. Value stream map
    C. SIPOC chart
    D. Swim lane mapping

157. What are some of the inputs to a process as listed in the text?
    A. Designs and measurements
    B. Needs, ideas, expectations, and requirements
    C. Products and services
    D. Decisions and authorizations

158. What is a SIPOC chart used for?
    A. To detail every step in a process
    B. To provide a high-level view of a process from suppliers to customers
    C. To identify defects in a product
    D. To train new employees

159. Which is NOT a category traditionally divided in a cause-and-effect diagram?
    A. Methods
    B. Machines
    C. Market trends
    D. Materials

160. Why is it important to update the flowchart when a change is made to the process?
    A. To make it look more professional
    B. To ensure compliance and accuracy
    C. To impress customers
    D. To add more colors and symbols

Practice Questions

161. What is the primary purpose of brainstorming in process analysis?
    A. To enforce ground rules.
    B. To encourage divergent thinking and identify the most possible causes.
    C. To prioritize ideas based on feasibility.
    D. To ensure team members do not criticize each other

162. Which tool is particularly effective for prioritizing ideas when the issue being discussed is sensitive and emotional?
    A. Affinity diagram
    B. Pareto chart
    C. Nominal Group Technique (NGT)
    D. Multi-voting

163. What does the Pareto chart help to visualize?
    A. The frequency of measurement errors.
    B. The relationship between different variables.
    C. The "vital few" and "trivial many" using the 80:20 principle.
    D. The detailed defect summary.

164. Which variation of brainstorming involves using sticky notes on a fishbone diagram displayed on a wall or board?
    A. Nominal Group Technique (NGT)
    B. Affinity Diagram
    C. CEDAC (cause-and-effect diagram with the addition of cards)
    D. Multi-voting

165. What is the primary purpose of the affinity diagram in brainstorming?
    A. To display the degree of the relationship between variables.
    B. To organize a large number of ideas into their natural relationships.
    C. To prioritize ideas based on their weighted score.

Practice Questions

D. To ensure data are not too specific.

166. What does a true Pareto chart include that makes it more informative than an ordinary bar graph?
    A. Data arranged in ascending order of occurrence.
    B. A secondary axis with percentages and a cumulative percentage line.
    C. Multiple colors to differentiate categories.
    D. A detailed defect summary with specifics as to location and appearance.

167. In brainstorming, what is a common tendency that should be avoided to ensure effective idea generation?
    A. Prioritizing ideas too quickly.
    B. Encouraging divergent thinking.
    C. Criticizing ideas instantly and discarding them during the session.
    D. Using multiple brainstorming variations.

168. What is the role of the facilitator in a brainstorming session?
    A. To enforce ground rules and encourage ideas.
    B. To prioritize ideas based on feasibility.
    C. To collect data on the various causes.
    D. To create a relationship matrix.

169. What is the primary purpose of the relationship diagram in process analysis?
    A. To display the degree of the relationship between variables.
    B. To visualize data arranged in descending order of frequency.
    C. To organize ideas into their natural relationships.
    D. To rank ideas based on redundancy and feasibility.

170. What should be avoided when performing a Pareto analysis?
    A. Arranging data in descending order of frequency.

Practice Questions

B. Pooling "trivial many" data as "miscellaneous" or "other."
C. Ensuring the data are not too specific.
D. Using cost as a measure for creating the Pareto chart.

171. What is the minimum sample size generally considered sufficiently large for the sampling distribution of the mean to be nearly normal?
    A. 10
    B. 20
    C. 30
    D. 40

172. What does the standard error of the mean represent?
    A. The standard deviation of the population
    B. The standard deviation of the sample
    C. The standard deviation of the sampling distribution of the mean
    D. The average of the sample means

173. Which of the following is true about the central limit theorem?
    A. It only applies to normally distributed populations
    B. It states that the sampling distribution of the mean will be normal regardless of the sample size
    C. It states that the sampling distribution of the mean will approximate a normal distribution as the sample size increases
    D. It only applies to sample sizes greater than 50

174. For a normally distributed population, approximately what percentage of data falls within ±1.96 standard deviations of the mean?
    A. 68%
    B. 95%
    C. 99%
    D. 99.7%

Practice Questions

175. If the sample mean is 10 milligrams and the standard error is 0.003, what is the 95% confidence interval for the mean?
    A. 9.9941 to 10.0059 milligrams
    B. 9.9000 to 10.1000 milligrams
    C. 9.9800 to 10.0200 milligrams
    D. 9.9700 to 10.0300 milligrams

176. Which type of statistical study uses data from a sample to make estimates or inferences about a population?
    A. Descriptive statistics
    B. Inferential statistics
    C. Enumerative statistics
    D. Graphical statistics

177. What is the primary purpose of control charts in the context of the central limit theorem?
    A. To measure individual data points
    B. To monitor the status of a process by averaging the measurement values in subgroups
    C. To display the full population data
    D. To compare the means of two different processes

178. If a sample of size 16 is drawn from a population with a historical standard deviation of 0.012 milligrams, what is the sample standard deviation?
    A. 0.012 milligrams
    B. 0.003 milligrams
    C. 0.048 milligrams
    D. 0.001 milligrams

179. What is the probability of obtaining exactly one defective item in a sample size of five, with a defective rate of 10%?

Practice Questions

A. 0.262
B. 0.328
C. 0.110
D. 0.450

180. Which of the following distributions is typically used to model the number of defects on a painted surface?
    A. Normal distribution
    B. Binomial distribution
    C. Poisson distribution
    D. Exponential distribution

181. What is the primary advantage of moving from discrete to continuous measurements?
    A. Continuous measurements require larger sample sizes
    B. Continuous measurements are less sensitive to process changes
    C. Continuous measurements require no additional infrastructure
    D. Continuous measurements are more sensitive to process changes

182. What is the key characteristic of continuous scales?
    A. They have a limited number of values
    B. They result from counting events
    C. They have infinite values between any two points
    D. They can only be used for nominal data

183. Which data collection method typically has low response rates?
    A. Face-to-face interviews
    B. Focus groups
    C. Surveys
    D. Mystery shopping

Practice Questions

184. What sampling method is used when there is no homogeneity in the lot being sampled?
    A. Random Sampling
    B. Sequential Sampling
    C. Stratified Sampling
    D. Simple Sampling

185. Why is real-time data acquisition preferred over manual data entry?
    A. It is always cheaper
    B. It eliminates the need for data coding
    C. It reduces human errors and allows for swift action
    D. It does not require additional infrastructure

186. What is an example of locational data?
    A. Number of defective valves
    B. Paint defects in an automobile assembly line
    C. Quantity of materials used
    D. Measurements of machine parts

187. What is a common cause of errors in data collection?
    A. Real-time data acquisition
    B. Multiple points of data entry
    C. Using continuous data
    D. Stratified sampling

188. Which arithmetic operation applies to ratio scales?
    A. Counting
    B. Addition and subtraction
    C. Multiplication and division
    D. Mode

189. What method can be used to minimize errors in manual data entry?
    A. Avoiding data coding

Practice Questions

   B. Conducting R&R studies
   C. Increasing sample sizes
   D. Using nominal scales

190. What is the purpose of a check sheet in data collection?
   A. To perform complex statistical analyses
   B. To pre-categorize potential outcomes for data collection
   C. To eliminate the need for manual data entry
   D. To measure continuous data only

191. What is the purpose of a check sheet in the context of quality improvement?
   A. To calculate complex statistical formulas
   B. To focus on continual improvement
   C. To generate random data for analysis
   D. To replace attribute statistical process control charts

192. Which of the following is NOT a measure of central tendency?
   A. Mean
   B. Median
   C. Mode
   D. Range

193. What does the sample standard deviation estimate?
   A. The mean of the sample
   B. The dispersion of the sample
   C. The mean of the population
   D. The standard deviation of the population

194. In a box plot, what does the upper whisker represent?
   A. The highest value within the upper limit
   B. The median of the dataset

Practice Questions

   C. The first quartile (Q1)
   D. The interquartile range (IQR)

195. What type of distribution is used to estimate the shape parameter b and mean time between failures (MTBF)?
   A. Normal distribution
   B. Binomial distribution
   C. Weibull distribution
   D. Poisson distribution

196. What does a scatter diagram help to identify?
   A. Measures of central tendency
   B. Frequency distributions
   C. Relationships or associations between two variables
   D. Cumulative frequency distributions

197. Which of the following tests is most widely used by statistical software to check the normality of random data?
   A. Ryan-Joiner test
   B. Kolmogorov-Smirnov test
   C. Anderson-Darling test
   D. Shapiro-Wilk test

198. How is the median value represented in a stem-and-leaf plot?
   A. With an asterisk (*)
   B. By bolding the value
   C. Enclosed in parentheses
   D. Underlined

199. What does a run chart primarily help to identify?
   A. Measures of central tendency
   B. Nonrandom patterns in process data

Copyright © 2024 VERSAtile Reads. All rights reserved.
This material is protected by copyright, any infringement will be dealt with legal and punitive action.

Practice Questions

    C. Measures of dispersion
    D. Frequency distributions

200. Which graphical method uses the high and low values of the data as well as the quartiles to present a pictorial view?
    A. Histogram
    B. Scatter diagram
    C. Stem-and-leaf plot
    D. Box-and-whisker plot

201. What is the primary purpose of Measurement System Analysis (MSA)?
    A. To increase production speed
    B. To reduce variation in the measurement system
    C. To improve customer satisfaction
    D. To lower production costs

202. Which of the following best describes repeatability in MSA?
    A. Variation in measurement when measured by two or more appraisers multiple times
    B. Variation in measurement when measured by one appraiser on the same equipment in the same measurement setting at the same time
    C. Drift in absolute value over time
    D. Accuracy of measurement at various measurement points of measuring range in the equipment

203. Which of the following is NOT a factor that influences reproducibility?
    A. Operator training, skill, and knowledge
    B. Consistency in measurement
    C. Design of the measurement system
    D. Special loading and unloading of the work piece

204. What is the first step in conducting a gage repeatability and reproducibility (GR&R) study?

Practice Questions

A. Create a table for experimentation
B. Call appraisers to perform measurements
C. Plan the study in detail
D. Select and identify samples for the GR&R study

205. What does the term 'bias' refer to in MSA?
A. The closeness of repeated readings to each other
B. The difference between an absolute value and the true value with respect to a standard master at various measurement points
C. Accuracy of measurement at various measurement points of measuring range in the equipment
D. Drift in absolute value over time

206. What is the significance of having appraisers measure all samples for every trial in a GR&R study?
A. To ensure that the study is complete and the calculations are accurate
B. To reduce the time taken to conduct the study
C. To minimize the cost of the study
D. To make the study more complex

207. Which of the following statements is true regarding the 'precision' in MSA?
A. It refers to the accuracy of measurement at various points of the measuring range
B. It refers to the closeness of repeated readings to each other
C. It refers to the drift in absolute value over time
D. It refers to the bias with respect to a standard master

208. In the context of MSA, what does 'linearity' refer to?
A. The drift in absolute value over time
B. The variation in measurement when measured by two or more appraisers multiple times
C. The closeness of repeated readings to each other

Practice Questions

D. The accuracy of measurement at various measurement points of the measuring range in the equipment

209. In a GR&R study, what should be done if the points in the sample range charts are outside the control limits?
A. Ignore the points and proceed with the study
B. Investigate to verify whether this is a recording error or due to any special causes
C. Discard the entire study and start over
D. Adjust the control limits to include the points

210. What does the 'EV' label stand for in the Gage Repeatability and Reproducibility Report?
A. Equipment Variation
B. Error Variation
C. External Variation
D. Environmental Variation

211. What does the AIAG MSA Manual describe as the "measurement resolution, scale limit, or smallest detectable unit of the measurement device and standard"?
A. Repeatability
B. Reproducibility
C. Discrimination
D. Linearity

212. What is considered an acceptable number of distinct categories for process monitoring applications?
A. Less than 1
B. Between 2 and 4
C. Greater than or equal to 5
D. Greater than 10

Practice Questions

213. Which source of variation is responsible for differences in measurements when different appraisers measure the same part?
    A. Repeatability
    B. Reproducibility
    C. Part-to-part
    D. Discrimination

214. What is the effect of high gage R&R on process capability (Cp) assessment?
    A. It decreases the error in Cp assessment
    B. It has no effect on Cp assessment
    C. It increases the error in Cp assessment
    D. It makes the Cp assessment more accurate

215. Which of the following is NOT a common error made while performing GR&R studies?
    A. Randomizing the samples during measurement
    B. Using untrained appraisers
    C. Publishing the results with appraisers' names
    D. Assuming the GR&R results are valid forever

216. What should be done to avoid introducing bias in repetitive measurement trials?
    A. Use the same appraiser for all trials
    B. Not randomize the samples
    C. Randomize the samples during measurement
    D. Use process-unrelated employees

217. What is the recommended approach when reporting GR&R results to avoid unhealthy comparisons between appraisers?
    A. Publish the results with appraisers' names
    B. Publish the results anonymously as appraiser A, B, C
    C. Only report the reproducibility errors

Practice Questions

   D. Do not publish the results at all

218. What should be done if the measurement equipment shows nonlinearity across the measurement range?
   A. Ignore the issue
   B. Calibrate the equipment properly at both high and low ends
   C. Use the equipment without changes
   D. Only calibrate the equipment at the midpoint

219. How often should GR&R results be validated?
   A. Once and assume they are valid forever
   B. Every year
   C. Periodically, just like gage calibration
   D. Only when there is a change in appraisers

220. What is the primary purpose of creating scatter diagrams and estimating the correlation coefficient between measurement systems?
   A. To measure part-to-part variation
   B. To identify bias in a single measurement system
   C. To compare multiple measurement devices
   D. To measure repeatability

221. What is the primary purpose of conducting ongoing measurement system analysis?
   A. To enhance the appearance of the measurement equipment
   B. To reduce the cost of measurement equipment
   C. To understand the uncertainty of the measurement system
   D. To increase production speed

222. In orthogonal regression analysis, what does it indicate if the intercept is close to 0 and the slope is close to 1?
   A. The two methods most likely provide equivalent measurements
   B. The two methods are significantly different

Practice Questions

C. There is a high level of equipment variance
D. The measurements are unreliable

223. In an attribute agreement analysis, which appraiser showed the highest within-appraiser assessment agreement?
    A. Appraiser A
    B. Appraiser B
    C. Appraiser C
    D. All appraisers showed the same level of agreement

224. Which appraiser had the highest agreement with the standard in the attribute agreement analysis?
    A. Appraiser A
    B. Appraiser B
    C. Appraiser C
    D. Appraisers B and C

225. What is indicated if the variance between measuring equipment shows a significant p-value in components of variance analysis?
    A. The equipment is functioning perfectly
    B. There is no need for further investigation
    C. It is an area for the experimenter to investigate
    D. The variance is negligible

226. In the context of measurement system analysis, what does GR&R stand for?
    A. Good Results & Reproducibility
    B. Gauge Repeatability & Reproducibility
    C. Great Reliability & Repeatability
    D. General Readiness & Reliability

227. Which of the following is NOT listed as a practical challenge in GR&R studies?

Practice Questions

    A. One-sided specification
    B. Skewed distribution
    C. Fully automated equipment
    D. High cost of raw materials

228. What is the significance of the 95% confidence interval in the context of measurement system analysis?
    A. It indicates the range within which the true value lies with 95% certainty
    B. It shows the exact value of the measurement
    C. It is used to measure the efficiency of the process
    D. It determines the cost-effectiveness of the measurement system

229. In the attribute agreement analysis, which appraiser is more consistent but also accepts the most bad parts?
    A. Appraiser A
    B. Appraiser B
    C. Appraiser C
    D. None of the appraisers accepted bad parts

230. What is the first action a Green Belt should take if a process is not in statistical control and does not meet specifications?
    A. Continue the process as normal
    B. Stop the process and immediately investigate
    C. Make minor adjustments and proceed
    D. Ignore the issue

231. What is the first step in conducting a process capability study?
    A. Identifying customer specifications
    B. Performing measurement system analysis
    C. Creating a control plan
    D. Calculating control limits

Practice Questions

232. Which of the following best describes assignable causes in a process?
    A. Natural variation in incoming material supply
    B. Interaction between process steps caused by the way processes are sequenced
    C. Significant changes in incoming material quality
    D. Manufacturing equipment design

233. What is the purpose of creating internal specifications tighter than customer specifications?
    A. To reduce costs
    B. To provide additional protection to customers
    C. To meet regulatory requirements
    D. To simplify the production process

234. What does the ZU value represent in process capability analysis?
    A. The number of standard deviations between the process average and the lower specification limit
    B. The number of standard deviations between the process average and the upper specification limit
    C. The average range of the process
    D. The average standard deviation of the process

235. In the context of control charts, what does it mean when a point is more than 3.00 standard deviations from the centerline?
    A. The process is in control
    B. The process is out of control
    C. The specification limits need to be adjusted
    D. The sample size is too small

236. Why is it important to ensure that the range chart is within control before reviewing the average chart?
    A. To meet customer specifications
    B. To ensure the variation is in control before making adjustments to

averages
C. To simplify the analysis
D. To reduce the cost of quality control

237. What is the natural process limit for a process with an average of 1.065 lbs and a standard deviation of 0.024 lbs?
A. 1.00 to 1.10 lbs
B. 0.993 to 1.137 lbs
C. 1.041 to 1.089 lbs
D. 1.050 to 1.080 lbs

238. What is a common reason for engineers to be hesitant about investigating out-of-control conditions?
A. Lack of training
B. Time investment
C. Insufficient data
D. Lack of management support

239. What does Test 4 in the X-bar chart indicate?
A. One point more than 3.00 standard deviations from centerline
B. 14 points in a row alternating up and down
C. Two out of three points more than two standard deviations from centerline
D. All points within 1.00 standard deviations from the centerline

240. What should be done before estimating process capability?
A. Identify customer specifications
B. Recalculate control limits
C. Perform a comprehensive process FMEA
D. Remove assignable (special) causes

241. What is the primary goal of a multi-vari study?
A. To maximize production output.

Practice Questions

B. To understand and minimize process variations.
C. To increase the temperature of the manufacturing process.
D. To train employees on new machinery.

242. Which of the following is NOT a type of variation analyzed in a multi-vari study?
   A. Cyclical variation
   B. Temporal variation
   C. Positional variation
   D. Seasonal variation

243. In a multi-vari study, what is positional variation also known as?
   A. Within-part variation
   B. Shift-to-shift variation
   C. Part-to-part variation
   D. Batch-to-batch variation

244. What type of variation occurs as a change over time?
   A. Positional variation
   B. Cyclical variation
   C. Temporal variation
   D. Random variation

245. Which of the following steps is part of the multi-vari sampling plan procedure?
   A. Ignoring sample size
   B. Recording time and values from each sample set in a table format
   C. Using only one sample
   D. Plotting a graph with measured values on the horizontal scale

246. What was the dominant type of variation identified in the stainless steel casting example before improvements were made?
   A. Within-part variation

Practice Questions

B. Top-to-bottom variation
C. Part-to-part variation
D. Shift-to-shift variation

247. Why was a new metal foundry sourced for the stainless steel casting process?
    A. To reduce within-part variation
    B. To reduce part-to-part variation
    C. To increase production speed
    D. To decrease material costs

248. What improvement was suggested to address the out-of-round condition in the stainless steel casting example?
    A. Increase the machining speed
    B. Change the material of the casting
    C. Improve pressure regulator control for the pneumatic chuck
    D. Pressure-wash the parts before burnishing

249. In the stainless steel casting example, what was identified as a potential cause for the remaining shift-to-shift variation?
    A. Chuck clamping pressure
    B. Tool wear
    C. Material inconsistencies
    D. Operator skill level

250. For a transactional process, what should be done before applying a multi-vari study to understand variation meaningfully?
    A. Increase the complexity of transactions
    B. Stratify the data by transaction type and complexity
    C. Reduce the number of transactions
    D. Ignore the differences in customer needs

Practice Questions

251. Which of the following statements is true about correlation and causation?
    A. Correlation always implies causation.
    B. Causation always implies correlation.
    C. Correlation never implies causation.
    D. Causation never implies correlation.

252. What is the range of the correlation coefficient, r?
    A. 0 to 1
    B. -1 to 0
    C. -1 to 1
    D. 0 to 2

253. If the correlation coefficient, r, is close to zero, what does it indicate?
    A. A strong positive relationship
    B. A strong negative relationship
    C. No linear relationship
    D. A perfect nonlinear relationship

254. In a scatter plot, how are the independent and dependent variables conventionally placed?
    A. Independent variable on the vertical axis and dependent variable on the horizontal axis
    B. Independent variable on the horizontal axis and dependent variable on the vertical axis
    C. Both variables on the horizontal axis
    D. Both variables on the vertical axis

255. What does a p-value less than 0.05 indicate in hypothesis testing?
    A. Accept the null hypothesis
    B. Reject the null hypothesis
    C. There is no statistical significance
    D. The result is inconclusive

Practice Questions

256. Which of the following represents a perfect positive correlation?
   A. r = 0
   B. r = -1
   C. r = 0.5
   D. r = 1

257. What is the equation for a linear regression line?
   A. y = mx + b
   B. y = mx^2 + b
   C. y = a + bx
   D. y = ax + b

258. What does the slope (b) in the regression equation, y = a + bx represents?
   A. The point where the line crosses the y-axis
   B. The predicted value of y for a given value of x
   C. The change in y for a one-unit change in x
   D. The average value of x

259. In the context of regression analysis, what does the standard error of the estimate (se) measure?
   A. The slope of the regression line
   B. The amount of dispersion of observed values around the regression line
   C. The y-intercept of the regression line
   D. The correlation coefficient

260. When using Excel to calculate the correlation coefficient, which function is used?
   A. =SUM
   B. =AVERAGE

Practice Questions

C. =CORREL
D. =STDEV

261. What is the purpose of including more than one independent variable in a regression model?
   A. To decrease the variation in the dependent variable (y)
   B. To increase the sampling error
   C. To explain a higher proportion of the variation in the dependent variable (y)
   D. To avoid the use of statistical software

262. In a simple linear regression, what is the equation used to represent the relationship between the dependent variable (y) and the independent variable (x)?
   A. $y = b_0 + b_1 x$
   B. $y = mx + c$
   C. $y = ax^2 + bx + c$
   D. $y = a + bx$

263. In hypothesis testing, what does the null hypothesis (H0) represent?
   A. The status quo or no effect or difference
   B. The alternative effect or difference
   C. The probability of making a Type I error
   D. The probability of making a Type II error

264. Which of the following describes a Type I error in hypothesis testing?
   A. Rejecting the null hypothesis when it is true
   B. Failing to reject the null hypothesis when it is false
   C. Accepting the null hypothesis when it is false
   D. Accepting the null hypothesis when it is true

Practice Questions

265. If a test statistic falls within the rejection region in a hypothesis test, what should you do?
    A. Accept the null hypothesis
    B. Fail to reject the null hypothesis
    C. Reject the null hypothesis
    D. Increase the sample size

266. What is the significance of a p-value in hypothesis testing?
    A. It represents the probability of making a Type II error
    B. It represents the probability of obtaining a test statistic at least as extreme as the one observed
    C. It is used to determine the sample size
    D. It represents the confidence interval

267. In a one-sample Z-test, if the population standard deviation is known, which formula is used to calculate the test statistic?
    A. $t = (X - m_0) / (s / \sqrt{n})$
    B. $Z = (X - m_0) / (s / \sqrt{n})$
    C. $t = (X - m_0) / s$
    D. $Z = (X - m_0) / s$

268. What is the formula for the confidence interval for the mean in a large sample (n ≥ 30)?
    A. $X \pm Z (s / \sqrt{n})$
    B. $X \pm t (s / \sqrt{n})$
    C. $X \pm Z (s)$
    D. $X \pm t (s)$

269. In a t-test, how is the test statistic calculated when the population variance is unknown?
    A. $Z = (X - m_0) / (s / \sqrt{n})$
    B. $t = (X - m_0) / (s / \sqrt{n})$

Copyright © 2024 VERSAtile Reads. All rights reserved.
This material is protected by copyright, any infringement will be dealt with legal and punitive action.

Practice Questions

C. Z = (X − mo) / s
D. t = (X − mo) / s

270. Which of the following is true about the t-distribution used in hypothesis testing?
A. It is always used for large samples
B. It is derived from a normal distribution with a known standard deviation
C. It applies to samples drawn from a normally distributed population with an unknown standard deviation
D. It does not depend on degrees of freedom

271. What is the null hypothesis in the chi-square test for variance given in the context?
A. σ-squared ≠ 36
B. σ-squared = 36
C. σ-squared > 36
D. σ-squared < 36

272. Based on the Minitab Analysis, what is the p-value for the chi-square test for variance?
A. 0.05
B. 0.257
C. 0.01
D. 0.005

273. What should be concluded if the p-value is greater than 0.05 in the chi-square test for variance?
A. Reject the null hypothesis
B. Accept the alternative hypothesis
C. Fail to reject the null hypothesis
D. The test is inconclusive

Practice Questions

274. What is the 95% confidence interval for the variance according to the Minitab Analysis?
A. (30.1, 79.0)
B. (5.49, 8.89)
C. (20.1, 60.0)
D. (10.1, 50.0)

275. Which method is used to determine the confidence interval for the variance?
A. Z-Distribution
B. T-Distribution
C. F-Distribution
D. Chi-Square Distribution

276. What is the sample variance of the 35 samples as found in the example calculation?
A. 46
B. 36
C. 50
D. 30

277. Which of the following conditions must be met for a chi-square test of variance to be valid?
A. The sample size must be greater than 50
B. The population must be normally distributed
C. The sample proportion must be less than 0.05
D. The sample mean must be zero

278. What is the chi-square statistic value for the variance test with 34 degrees of freedom?
A. 35.0
B. 43.44
C. 30.1
D. 46.0

Practice Questions

279. If the new process variance must be either lower than 30 or higher than 79 to declare a difference, what does the current analysis conclude?
    A. There is a significant difference
    B. There is no significant difference
    C. There is not enough information to conclude
    D. The test must be redone

280. What is the standard deviation range at a 95% confidence interval according to the Minitab Analysis?
    A. (5.49, 8.89)
    B. (5.00, 9.00)
    C. (4.50, 8.50)
    D. (6.00, 9.00)

281. What term refers to the settings or possible values of a factor in an experimental design?
    A. Factors
    B. Levels
    C. Treatments
    D. Blocks

282. What is the main purpose of replication in DOE?
    A. To measure the same treatment multiple times under the same setup
    B. To reduce the overall cost of the experiment
    C. To increase precision and balance unknown factors
    D. To organize the experiment in a chance manner

283. Which of the following is a variable that depends on another variable in an experiment?
    A. Independent variable

Practice Questions

   B. Factor
   C. Dependent variable
   D. Treatment

284. What is the function of randomization in DOE?
   A. To reduce the time taken to complete the experiment
   B. To minimize the effect of unwanted variables
   C. To organize treatment combinations in a chance manner
   D. To ensure treatments are assigned to all possible levels

285. What term describes the portion of experimental material or environment that is common to itself and distinct from other portions?
   A. Block
   B. Factor
   C. Treatment
   D. Level

286. What is the main difference between replication and repetition in DOE?
   A. Replication involves multiple factors, while repetition involves a single factor
   B. Replication is done in random order, while repetition is done under the same setup
   C. Replication reduces costs, while repetition increases costs
   D. Replication affects dependent variables, while repetition affects independent variables

287. Which of the following best describes an experimental error?
   A. A planned factor level adjustment
   B. Variation in the response variable when levels and factors are held constant
   C. The process of repeating experimental trials

Practice Questions

   D. The interaction between two independent variables

288. What does the term 'treatment combination' refer to in DOE?
   A. The interaction between two factors
   B. A single level assigned to a single factor
   C. The set of levels for all factors in a given experimental run
   D. The random assignment of levels to factors

289. What is the primary objective of a designed experiment in DOE?
   A. To minimize experimental costs
   B. To generate knowledge about a product or process
   C. To randomize all experimental runs
   D. To ensure all factors are held constant

290. Which term describes the outcome of an experiment as a result of controlling the levels, interactions, and number of factors?
   A. Factor
   B. Response
   C. Treatment
   D. Block

291. What is the primary purpose of a completely randomized experiment?
   A. To analyze multiple factors simultaneously
   B. To investigate a single factor
   C. To reduce the number of experimental runs
   D. To create contour diagrams of factors' influence

292. Which type of design is used when investigating several factors at two or more levels and interactions are necessary?
   A. Blocked factorials
   B. Fractional factorials
   C. Factorials
   D. Randomized blocks

Practice Questions

293. What is a response variable?
    A. A factor that is controlled
    B. An external variable that correlates with other variables
    C. A variable influenced by other factors
    D. A variable that affects the other factors

294. What is the primary purpose of randomization in DOE?
    A. To eliminate variation
    B. To lessen the effects of special cause variation
    C. To increase the number of experiments
    D. To ensure all factors are tested equally

295. Which design should be chosen if the experiment requires a larger number of blocks?
    A. Balanced incomplete blocks
    B. Completely randomized design
    C. Partially balanced incomplete blocks
    D. Latin square

296. What should be the first consideration when preparing to experiment?
    A. The available budget
    B. The number of factors
    C. The experimental objective
    D. The measurement system

297. In a Latin square design, what must be equal?
    A. The number of blocks
    B. The number of rows, columns, and treatments
    C. The levels of factors
    D. The number of experiments

Practice Questions

298. What is a mixture design?
    A. A design that investigates a single factor
    B. A factorial experiment with factor levels expressed in percentages
    C. A design that reduces the combinations of factors and levels
    D. A design that uses blocking to run experiments in subsets

299. Which experimental design is appropriate for studying relative variability instead of mean effects?
    A. Fractional factorials
    B. Nested designs
    C. Randomized blocks
    D. Latin square

300. Which design is used to control the interactions and effects of other variables while investigating a primary factor?
    A. Factorial design
    B. Mixture design
    C. Latin square
    D. Completely randomized design

301. What is the primary purpose of using a Design of Experiments (DOE)?
    A. To solve all operational problems
    B. To understand the variation in a process through testing and optimizing
    C. To reduce the production costs
    D. To increase the number of variables in a process

302. Which of the following is a method that attempts to lessen the effects of special cause variation by grouping the experiments in batches?
    A. Replication
    B. Screening
    C. Blocking

Practice Questions

   D. Confounding

303. What term is used when one or more factors influence the behavior of another factor?
   A. Replication
   B. Interaction
   C. Confounding
   D. Screening

304. In a randomized block design, how might you nullify the impact of a shift change?
   A. By conducting all experiments during the first shift
   B. By doing the first three replicates of each run during the first shift and the remaining two replicates during the second shift
   C. By conducting all experiments during the second shift
   D. By randomly selecting shifts for each experiment

305. Which of the following represents a balanced experimental design?
   A. Each setting of each factor appears a different number of times with each setting of every other factor
   B. Each setting of each factor appears the same number of times with each setting of every other factor
   C. The factors are tested in a non-random order
   D. Only one factor is varied at a time

306. What is the purpose of conducting confirmation tests in DOE?
   A. To increase the complexity of the experiment
   B. To determine if the experiment has successfully identified better process conditions
   C. To reduce the number of trials
   D. To ensure the experiment is balanced

307. What is the term used when a factor interaction cannot be separately determined from a major factor in an experiment?

Practice Questions

A. Replication
B. Screening
C. Confounding
D. Interaction

308. Which step in DOE involves brainstorming the key factors causing variation?
    A. Defining the experiment
    B. Collecting samples from each run
    C. Mapping the current process
    D. Brainstorming the causes of variation

309. In a completely randomized design, what is true about the order of tests?
    A. The tests are conducted in a predetermined order
    B. No tests are omitted, and the order is completely random
    C. The tests are grouped by factors
    D. The order is based on the level of difficulty

310. What is the significance of using ANOVA procedures in DOE?
    A. To increase the number of replicates
    B. To determine whether the difference between high and low values are due to real difference or experimental error
    C. To randomize the order of experiments
    D. To ensure each factor appears the same number of times

311. In a full factorial experiment, what is the formula to determine the number of runs?
    A. Number of runs = L + F
    B. Number of runs = L * F
    C. Number of runs = L^F
    D. Number of runs = L / F

Practice Questions

312. Which factor had the greatest effect on the steak cooking approval rating in the full factorial example?
    A. Cooking method
    B. Meat type
    C. Marinade
    D. Interaction between meat and marinade

313. Which of the following methodologies is primarily used for root cause analysis?
    A. Full factorial design
    B. Fractional factorial design
    C. Failure Mode and Effects Analysis (FMEA)
    D. Fishbone Diagram

314. What is the purpose of a SIPOC diagram in Six Sigma?
    A. To create a visual representation of the problem
    B. To identify potential failure modes
    C. To define the boundaries and scope of a process
    D. To collect customer feedback

315. What is the primary advantage of conducting a fractional factorial experiment over a full factorial experiment?
    A. It examines every possible combination
    B. It saves time and resources
    C. It is more accurate
    D. It provides more data

316. When analyzing the data from a two-level fractional factorial experiment, which statistical method is appropriate for assessing significance?
    A. Linear Regression
    B. ANOVA (Analysis of Variance)

Practice Questions

   C. Chi-Square Test
   D. T-Test

317. Which of the following is NOT a typical phase in the root cause analysis methodology?
   A. Define and document the problem
   B. Collect and analyze data
   C. Implement a temporary solution
   D. Establish a corrective action plan

318. In the described corrective action framework, which phase involves creating or updating control plans and procedures?
   A. Identify the Opportunity
   B. Analyze the Current Process
   C. Implement Changes
   D. Standardize the Solution

319. What is the purpose of the 5 Whys method in root cause analysis?
   A. To list all possible causes
   B. To prioritize potential solutions
   C. To drive to a deeper understanding of the cause
   D. To document the problem

320. In the context of root cause analysis, what does a cause-and-effect (X-Y) relational matrix help to achieve?
   A. It lists all potential causes
   B. It visually represents the problem
   C. It quantifies and prioritizes the impacts of causes on effects
   D. It provides a timeline for implementing solutions

321. Which of the following is NOT a component of the 5S methodology?
   A. Sort
   B. Standardize

Practice Questions

   C. Simplify
   D. Sustain

322. What is the primary objective of implementing a pull system in Lean?
   A. Increase work-in-process inventory
   B. Reduce overproduction
   C. Maximize lead times
   D. Complicate scheduling

323. What does the term 'poka-yoke' refer to in Lean practices?
   A. Continuous improvement events
   B. Mistake-proofing methods
   C. Inventory management
   D. Visual control systems

324. How does Lean differ from traditional systems in terms of inventory?
   A. Lean increases work-in-process inventory
   B. Lean reduces the need for large, segregated inventory areas
   C. Lean creates more inventory for safety stock
   D. Lean does not consider inventory management

325. Which tool is used in Lean to create order and organization in the workspace?
   A. Kanban
   B. 5S
   C. Kaizen
   D. SIPOC

326. What is the goal of Total Productive Maintenance (TPM) in Lean?
   A. Increase setup times
   B. Manage operations with preventive maintenance
   C. Increase machine breakdowns
   D. Complicate process flow

Practice Questions

327. Which Lean tool involves visual displays and controls to show the status of processes?
    A. Visual Factory
    B. Kanban
    C. Poka-yoke
    D. Kaizen

328. What is a Kaizen event primarily focused on?
    A. Large-scale, expensive improvements
    B. Low-cost, gradual improvement of processes
    C. Increasing inventory levels
    D. Reducing communication

329. What is a key benefit of standard work in Lean?
    A. Increases process variation
    B. Provides a baseline for improvement
    C. Reduces process stability
    D. Eliminates the need for training

330. What does process capability represent in lean thinking?
    A. The speed of the process
    B. A comparison of the voice of the customer (VOC) with the voice of the process (VOP)
    C. The cost of production
    D. The total cycle time

331. Which tool is used to calculate and understand the capacity of equipment in a particular work area?
    A. Value Stream Mapping
    B. Process Capacity Sheet
    C. Takt Time Analysis
    D. Flowchart

Practice Questions

332. In the context of lean, what does work flow analysis evaluate?
   A. The cost of materials
   B. The layout of the equipment
   C. The flow of work through a system
   D. The number of employees required

333. What is the primary purpose of takt time in lean manufacturing?
   A. To measure the total production cost
   B. To calculate the total cycle time
   C. To serve as a management tool to align production with customer demand
   D. To determine the number of employees required for a task

334. What is the goal of one-piece flow in lean manufacturing?
   A. To reduce the number of employees required
   B. To produce large batches of products
   C. To make one part at a time correctly and without unplanned interruptions
   D. To increase the time taken for each process step

335. What does the theory of constraints (TOC) approach focus on?
   A. Increasing the total cycle time
   B. Reducing the number of employees
   C. Identifying and managing bottlenecks in the process
   D. Increasing the inventory levels

336. Which lean tool allows for the identification of value-added and non-value-added activities?
   A. Process Capacity Sheet
   B. Flow Analysis Chart
   C. Value Stream Mapping
   D. Takt Time Analysis

Practice Questions

337. What is the purpose of quick changeover/setup reduction in lean manufacturing?
    A. To increase the total production time
    B. To reduce the time taken to switch from producing one product to another
    C. To increase the number of employees required
    D. To increase the inventory levels

338. In lean thinking, what does kanban refer to?
    A. A process for reducing cycle time
    B. A visual signal for replenishment of materials
    C. A method for mistake-proofing
    D. A type of flowchart

339. What is the primary objective of Kaizen?
    A. To achieve major redesigns of business processes
    B. To promote small, incremental process improvements
    C. To implement new technology
    D. To replace long-term projects with single large projects

340. What distinguishes a Kaizen Blitz from traditional Kaizen?
    A. It focuses on long-term, incremental improvements
    B. It involves major changes over multiple years
    C. It completes a continuous project within a week
    D. It replaces Kaizen efforts entirely

341. Which of the following is NOT a characteristic of a Constraint Resource (CCR)?
    A. It can compromise throughput if neglected
    B. It affects processes even if not running at full capacity
    C. It is the same as a bottleneck
    D. Examples include personnel, equipment, and suppliers

Practice Questions

342. What is the main focus of Gemba Kaizen?
    A. Major organizational changes
    B. Conducting improvements away from the workplace
    C. Improvement at the actual place where work happens
    D. Introducing new technology

343. Which of the following best describes Kaikaku?
    A. Evolutionary small changes
    B. Continuous small improvements
    C. Major revolutionary changes
    D. Minor adjustments to existing processes

344. What is the purpose of rational subgrouping in control charts?
    A. To enhance randomness and reduce piece-to-piece variation
    B. To increase sample size for more data
    C. To eliminate the need for control limits
    D. To use only the most critical dimensions of the product

345. What is the primary goal of Statistical Process Control (SPC)?
    A. To create arrays of control charts
    B. To monitor and improve process performance
    C. To replace manual processes with automated ones
    D. To document procedures without using data

346. What does a state of statistical control mean?
    A. Products meet specifications
    B. Only random causes are present in the process
    C. All special causes are present
    D. Control limits are not used

347. Which of the following is NOT a typical attribute contributing to Kaizen success?
    A. Presenting results
    B. Management commitment

Practice Questions

C. Identification of critical success factors
D. Ignoring follow-through

348. What does the term "Gemba" specifically refer to?
    A. A theoretical improvement model
    B. The place where real work happens
    C. The management office
    D. A type of statistical process control

349. What is referred to as under-adjustment or under-control in a process?
    A. Over-adjustment
    B. Responding to special causes of variation
    C. Failing to respond to special causes of variation
    D. Using control charts

350. Which type of data requires the use of control charts that incorporate measurements of central tendency or variation?
    A. Discrete data
    B. Attributes data
    C. Continuous data
    D. Binary data

351. What is the primary objective of Six Sigma?
    A. To increase a company's profit margins
    B. To improve leadership skills within an organization
    C. To reach a nearly perfect quality level with 3.4 defects per million opportunities
    D. To expand the company into new markets

352. Which of the following is considered a tool in quality management?
    A. Statistical process control
    B. Benchmarking

Practice Questions

   C. Failure mode and effects analysis
   D. Histogram

353. What is the main purpose of the Define phase in DMAICS?
   A. To identify key process characteristics
   B. To pinpoint the root causes of the problem
   C. To form a project team and draft a project charter
   D. To measure the performance of the process

354. In a Force Field Analysis, what are the forces that support change called?
   A. Restraining forces
   B. Driving forces
   C. Neutral forces
   D. Compelling forces

355. Which statistical measure is used to determine the inherent reproducibility of a process?
   A. Mean
   B. Standard Deviation
   C. Process Capability Index (Cp)
   D. Control Limits

356. Is it mandatory to undergo Lean Six Sigma training before taking the Green Belt Certification Exam?
   A. Yes, it is strictly required
   B. No, but it is recommended
   C. Yes, but only through online courses
   D. No, and it is not recommended

Practice Questions

357. Which principle is emphasized in the application of the Interrelationship Diagram?
    A. Linear thinking
    B. Multiple directional thinking
    C. Single-cause analysis
    D. Numerical data analysis

358. What is the primary focus of the Six Sigma methodology?
    A. Reducing cycle times
    B. Increasing employee satisfaction
    C. Reducing defects and variation
    D. Enhancing marketing efforts

359. What is the primary objective of the Improve phase in a project?
    A. To identify areas for improvement
    B. To develop and validate solutions
    C. To control the new process
    D. To define the project scope

360. What is the main benefit of the 'Control' phase in DMAICS?
    A. It identifies the cause of the problem
    B. It ensures the improved level of performance is consolidated
    C. It develops design alternatives
    D. It measures the key process characteristics

361. What is the primary goal of Green Six Sigma?
    A. Achieving short-term financial gains
    B. Enhancing product aesthetics
    C. Achieving sustainable results
    D. Minimizing employee training

Practice Questions

362. What is the primary goal of Design for Six Sigma (DFSS)?
    A. To fix defects in existing products
    B. To design products and processes that meet customer needs from the beginning
    C. To reduce production costs
    D. To enhance marketing strategies

363. What does DMAIC stand for in the context of Six Sigma?
    A. Design, Manufacture, Assure, Inspect, Control
    B. Define, Measure, Analyze, Improve, Control
    C. Develop, Manage, Activate, Implement, Check
    D. Discover, Map, Assess, Integrate, Confirm

364. Which tool is primarily used in the Measure phase to identify fail points in a process?
    A. SIPOC diagram
    B. Flow diagram
    C. Cause and effect diagram
    D. Histogram

365. Which of the following is NOT a barrier to achieving and sustaining results in quality initiatives, according to Basu (2001)?
    A. Packaged approach of the program
    B. High start-up costs
    C. Frequent changes in technology
    D. Top-down directive

366. Which tool is also known as the KJ method?
    A. Nominal group technique
    B. Affinity diagram
    C. SMED
    D. Value stream mapping

Practice Questions

367. What does the acronym OEE stand for in the context of Six Sigma?
    A. Overall Equipment Evaluation
    B. Overall Efficiency Effectiveness
    C. Overall Equipment Effectiveness
    D. Overall Equipment Efficiency

368. Which certification does IASSC provide for Green Belts?
    A. ICBB™
    B. ICGB™
    C. ICCYB™
    D. ICMBB™

369. Which chart is also known as a bar chart?
    A. Radar Chart
    B. Gantt Chart
    C. PDCA Cycle
    D. Milestone Tracker Diagram

370. In a control chart, what does a point above the Upper Control Limit (UCL) indicate?
    A. Common cause variation
    B. No variation
    C. Special cause variation
    D. Normal operation

371. What percentage of their time do Lean Six Sigma Green Belts typically spend on Lean Six Sigma projects?
    A. 50-75%
    B. 10-25%
    C. 30-50%

Practice Questions

D. 75-100%

372. What is the primary objective of the Define phase in the DMAICS cycle?
A. To measure key process characteristics
B. To identify, prioritize, and select the right project
C. To develop design alternatives
D. To control the improved process

373. What term is synonymous with 'FIT' in the context of quality processes?
A. Six Sigma
B. DMAIC
C. FIT SIGMA
D. Lean Sigma

374. Which technology is described as 'trustware' due to its ability to enhance transparency and traceability?
A. IoT
B. Cloud computing
C. AI
D. Blockchain

375. Which term is used to describe the ability of a process to produce a product that meets specifications?
A. Process variability
B. Process capability
C. Process alignment
D. Process control

Practice Questions

376. What does DMAIC stand for in Six Sigma?
    A. Define, Measure, Analyze, Improve, Control
    B. Define, Measure, Assess, Improve, Control
    C. Design, Measure, Analyze, Improve, Control
    D. Define, Measure, Analyze, Implement, Control

377. Which of the following is NOT a tool used in the Measure phase?
    A. Control charts
    B. Scatter diagram
    C. Check sheets
    D. Flow diagram

378. What is the purpose of mistake proofing?
    A. To increase production speed
    B. To prevent errors from becoming defects
    C. To reduce material costs
    D. To improve employee training

379. What is the primary focus of a PESTLE analysis?
    A. Internal process improvements
    B. External factors affecting a project
    C. Financial performance
    D. Team building exercises

380. Which of the following is NOT a phase outlined by the IASSC Lean Six Sigma Green Belt Body of Knowledge?
    A. Define
    B. Measure
    C. Execute
    D. Improve

Practice Questions

381. What does EVM stand for in project management?
    A. Earned Value Measurement
    B. Estimated Value Management
    C. Earned Value Management
    D. Efficient Value Measurement

382. What are the three cornerstones of the FIT methodology?
    A. Quality tools, process mapping, training
    B. Fitness for purpose, Sigma for improvement and integration, fitness for sustainability
    C. Lean principles, Six Sigma tools, DMAIC
    D. Customer focus, process improvement, cost reduction

383. Which of the following is NOT listed as an urgently required improved technology for climate change initiatives?
    A. Zero-carbon cement
    B. Carbon capture and storage
    C. Non-renewable energy
    D. Plant-based meat and dairy

384. Which tool uses a plot of points to study the relationship between two variables?
    A. Pareto Chart
    B. Scatter Diagram
    C. Histogram
    D. Control Chart

385. What is the Risk Priority Number (RPN) used in FMEA?
    A. The average number of defects per million opportunities
    B. The product of severity, occurrence, and detection ratings
    C. The sum of severity, occurrence, and detection ratings
    D. The difference between the severity and occurrence ratings

Practice Questions

386. How many questions does the IASSC Certified Lean Six Sigma Green Belt Exam™ consist of?
    A. 50 questions
    B. 100 questions
    C. 150 questions
    D. 200 questions

387. How much has the UK Government allocated for its flood protection program announced in July 2020?
    A. $3 billion
    B. $4.5 billion
    C. $6.2 billion
    D. $10 billion

388. What initiative does Bill Gates suggest to help farmers adapt to climate change?
    A. CGIAR (Consultative Group for International Agricultural Research)
    B. Increased fossil fuel usage
    C. Urban development
    D. Reduced crop variety

389. Which tool helps identify the most critical tasks that might delay a project?
    A. Radar Chart
    B. Activity Network Diagram
    C. Gantt Chart
    D. PDCA Cycle

Practice Questions

390. Which phase follows the Improve phase in a Six Sigma project?
    A. Define
    B. Measure
    C. Analyze
    D. Control

391. What is the main cause of the reduction in fatalities during Cyclone Amphan in Bangladesh?
    A. Advanced technology
    B. Early warning systems and concrete cyclone shelters
    C. International isolation
    D. Increased population

392. What is the purpose of the Analyse phase in DMAICS?
    A. To identify the right project
    B. To implement solutions
    C. To pinpoint the root causes of the problem
    D. To measure key process characteristics

393. Which phase in DMAICS focuses on sustaining the improved process and environmental standards?
    A. Define
    B. Measure
    C. Control
    D. Sustain

394. Which of the following is NOT an example of a climate adaptation measure for houses?
    A. Installing solar panels
    B. Using fossil fuel heaters
    C. Applying a circular economy
    D. Installing quick-fit flood-barriers

Practice Questions

395. Which company is known as the pioneer of Six Sigma?
    A. General Electric
    B. AlliedSignal
    C. Motorola
    D. Bechtel

396. What is the minimum passing score for the IASSC Certified Lean Six Sigma Green Belt Exam?
    A. 50%
    B. 60%
    C. 70%
    D. 80%

397. What does Green Six Sigma primarily focus on in the context of supply chains?
    A. Increasing waste
    B. Reducing efficiency
    C. Reducing waste and increasing efficiency
    D. Promoting fossil fuels

398. What is a significant outcome of Walmart's supply chain expertise during Hurricane Katrina?
    A. Faster and more effective disaster relief than FEMA
    B. Increased profits
    C. Development of new product lines
    D. Expansion into new markets

399. Which quality dimension does David Garvin's model NOT include?
    A. Reliability
    B. Durability
    C. Serviceability

Practice Questions

D. Cost reduction

400. Which company applied Six Sigma to improve the operations of its subsidiary, FilmTec Corp?
A. General Electric
B. Dow Chemical
C. Seagate Technology
D. Bechtel

401. Which economic sector is the largest contributor to greenhouse gas emissions?
A. Agriculture and land
B. Industry
C. Electricity and other energy
D. Transports

402. What is the final step in the basic process mapping procedure?
A. Simulate what should be happening
B. Construct a process diagram
C. Apply 'what if' analysis and simulation
D. Allocate ownership of each activity

403. What is a key aspect of Total Quality Management according to ISO 8402:1995?
A. Focus on short-term success
B. Management approach centered on quality
C. Emphasis on individual achievements
D. Exclusive use of statistical methods

404. What is the first step in creating a Gantt chart?
A. Draw vertical lines to divide the board.

Practice Questions

   B. Identify key activities or tasks.
   C. Estimate start and finish dates.
   D. Arrange activities in sequence

405. Which technique is commonly associated with the term 'random walk' in its problem-solving method?
   A. Design of Experiments (DOE)
   B. TRIZ
   C. Monte Carlo Technique (MCT)
   D. Statistical Process Control (SPC)

406. What does the acronym SMED stand for?
   A. Single Minute Exchange of Dies
   B. Six-Minute Error Detection
   C. Standardized Method for Efficiency Design
   D. Simple Manufacturing Error Detection

407. What is the primary goal of the Analyze phase in DMAICS?
   A. To identify the problem
   B. To measure the problem
   C. To pinpoint the root causes of the problem
   D. To implement solutions

408. What feature is mentioned as part of the architectural design to adapt houses to heatwaves?
   A. Roof insulation
   B. Cavity wall insulation
   C. Smart glass windows
   D. Reinforced roofing

Practice Questions

409. Which professional designation is granted upon successful completion of the Green Belt Certification?
    A. IASSC Certified Master Black Belt
    B. IASSC Certified Black Belt
    C. IASSC Certified Yellow Belt
    D. IASSC Certified Green Belt

410. Which tool is suggested for use in accelerating R&D projects on alternative fuels?
    A. Kaizen
    B. DFSS (Design for Six SigmA)
    C. 5S
    D. Gemba Walk

411. What is the target year for achieving zero marine plastic litter under the Osaka Blue Ocean Vision plan?
    A. 2030
    B. 2040
    C. 2050
    D. 2060

412. Which tool is used in the Sustain phase to ensure a continuous improvement culture?
    A. Run chart
    B. Pareto chart
    C. Balanced Scorecard
    D. Histogram

413. What type of diagram would you use to represent a physical process linked to the sequence of operations?
    A. Flow Process Chart
    B. Run Chart

Practice Questions

   C. Pareto Chart
   D. Scatter Diagram

414. According to Juran, quality is primarily achieved through:
   A. Post-production inspection
   B. Continuous education and training
   C. Reducing the cost of production
   D. Implementing more supervision

415. What is the purpose of the "Bring Back Mangroves Project"?
   A. To clear mangrove forests for buildings
   B. To support fisheries and protect coastlines
   C. To promote urban development
   D. To reduce tourism in coastal areas

416. Which phase of DMAICS involves the use of control charts to monitor the improved process?
   A. Define
   B. Measure
   C. Improve
   D. Control

417. Which tool is used in the Improve phase to generate ideas and solutions to solve problems?
   A. Histogram
   B. Brainstorming
   C. SIPOC diagram
   D. Control chart

418. What kind of project roles are Six Sigma Green Belts prepared to undertake?
   A. Team members on more complex projects
   B. Leading small, less critical projects

Practice Questions

C. Both A and B
D. Neither A nor B

419. What key concept is emphasized by Green Six Sigma that is not typically a focus in traditional Six Sigma?
A. Cost reduction
B. Customer satisfaction
C. Environmental sustainability
D. Employee training

420. Which country is mentioned as having successfully implemented a layered adaptation plan for cyclones?
A. India
B. Bangladesh
C. Indonesia
D. Philippines

421. Which of the following is NOT a measure for climate adaptation in retrofitting houses?
A. Reduce energy losses and consumption
B. Replace fossil fuel boilers and water heaters
C. Use non-renewable energy
D. Apply a circular economy

422. What is the main focus of Statistical Process Control (SPC)??
A. Cost reduction
B. Quality improvement through statistical methods
C. Employee motivation
D. Market analysis

423. What strategy does Quality 4.0 emphasize?

Practice Questions

   A. Increased manual inspections
   B. Digital tools and real-time data
   C. Traditional quality control
   D. Decreased use of technology

424. Which tool is used to categorize verbal information into an organized visual pattern?
   A. Brainstorming
   B. Mind mapping
   C. Affinity diagram
   D. Fishbone diagram

425. What is the primary benefit of using mobile technology in business operations?
   A. Decreased productivity
   B. Increased operational costs
   C. Enhanced communication and data access
   D. Reduced data security

426. What percentage of greenhouse gases is attributed to the transportation industry?
   A. 5%
   B. 10%
   C. 14%
   D. 20%

427. What is a Gantt chart primarily used for in the Control phase?
   A. To measure process performance
   B. To identify root causes
   C. To monitor project schedules
   D. To document project charters

Practice Questions

428. Which measure indicates the reliability of process tolerances to deliver customer specifications?
A. Defects per Million Opportunities (DPMO)
B. Process Capability Index (Cp)
C. Mean
D. Standard Deviation

429. What is a significant benefit of using Earned Value Management (EVM)?
A. It guarantees project success.
B. It provides real-time project tracking.
C. It enables variance analysis and forecasting.
D. It eliminates the need for project reviews.

430. How much does it cost to purchase a Green Belt Exam Voucher?
A. USD 150
B. USD 250
C. USD 350
D. USD 450

431. What is the primary goal of Unilever's climate adaptation strategy mentioned in the text?
A. Eliminate all carbon emissions by 2039
B. Invest €1 billion in landscape restoration over 10 years
C. Reduce the price of ice cream
D. Increase plastic usage

432. What innovative fuel is mentioned as reducing carbon emissions by 65% compared to fossil fuels?
A. Pure hydrogen
B. Methanol

Practice Questions

C. Hybrid fuel of 75% methanol and 25% hydrogen
D. Natural gas

433. What additional phase is included in Green Six Sigma compared to traditional Six Sigma?
    A. Design
    B. Sustain
    C. Verify
    D. Implement

434. What is a key focus in adapting green supply chains to climate change?
    A. Increasing fossil fuel usage
    B. Reducing renewable energy investment
    C. Enhancing resilience to adapt to new business conditions
    D. Eliminating international trade

435. Which technique involves the analysis of 'special' and 'common' causes of variation?
    A. Failure Mode and Effects Analysis (FMEA)
    B. Statistical Process Control (SPC)
    C. Quality Function Deployment (QFD)
    D. Design of Experiments (DOE)

436. Which technique is known for allowing quieter team members to have the same influence as more dominant ones?
    A. Affinity diagram
    B. Nominal group technique
    C. SMED
    D. Mind mapping

Practice Questions

437. Which of the following is NOT a component of the FIT methodology?
    A. Fitness for purpose
    B. Sigma for improvement and integration
    C. Quality assurance
    D. Fitness for sustainability

438. What is the main advantage of developing a company's own assessment process for quality initiatives?
    A. It is cheaper than hiring consultants
    B. It is always more accurate
    C. It ensures the company is aware of requirements and criteria
    D. It requires less training

439. Which methodology is mentioned as being modified for more effective research in carbon capture technology?
    A. DMAIC
    B. DMADV
    C. PDCA
    D. Six Sigma Black Belt

440. According to the text, what is the world's "last best chance" to reboot climate change targets and initiatives?
    A. The Paris Climate Agreement
    B. United Nations COP26 climate conference
    C. Kyoto Protocol
    D. Earth Day 2021

441. Lean manufacturing is characterized by:
    A. High levels of inventory
    B. Just-in-time production
    C. Extensive use of buffer stocks

Practice Questions

   D. Long production runs

442. What is the fundamental strategy of climate change preparation, as discussed in the text?
   A. Reduce, Reuse, Recycle
   B. Predict, Prevent, Preserve
   C. Stop, Think, Act
   D. Monitor, Control, Adapt

443. What does a scatter diagram NOT show?
   A. Positive correlation
   B. Negative correlation
   C. No correlation
   D. Exact cause and effect relationship

444. Which of the following is NOT a characteristic of Quality 4.0?
   A. Use of big data analytics
   B. Manual data collection
   C. Implementation of AI
   D. Integration of blockchain technology

445. What is a common, inexpensive method to prevent draughts in homes?
   A. Installing single-pane windows
   B. Using self-adhesive foam strips for doors and windows
   C. Keeping windows open during winter
   D. Using ceiling fans

446. Which type of boiler is suitable for a larger home with a greater demand for hot water?
   A. Combi boiler

B. Heat-only boiler
C. Electric boiler
D. Oil boiler

447. Which method is most commonly used for constructing an activity network diagram?
A. Fishbone method
B. Activity on arrow method
C. Pareto method
D. Histogram method

448. Which of the following is a principle of Total Productive Maintenance (TPM)?
A. Focus on individual goals
B. Six Sigma certification
C. Improvement of manufacturing efficiency
D. High batch production

449. What is the purpose of the initial assessment in FIT SIGMA?
A. To finalize the deployment plan
B. To evaluate the current process performance
C. To identify the real requirements before starting the program
D. To train employees on Six Sigma tools

450. What is the first step in creating an affinity diagram?
A. Collect current data on the problem
B. Clarify the chosen problem in a full sentence
C. Sort ideas into related groups
D. Label header cards

Practice Questions

451. What was the reported savings for GE from Six Sigma in 1999?
    A. $600 million
    B. $1 billion
    C. $2 billion
    D. $3 billion

452. Which phase of DMAICS involves brainstorming solutions to solve the problem?
    A. Define
    B. Improve
    C. Measure
    D. Control

453. Which of the following is NOT a characteristic assessed in Measurement System Analysis (MSA)?
    A. Bias
    B. Linearity
    C. Stability
    D. Profitability

454. What is the most common insulation material recommended for floorboards?
    A. Fiberglass
    B. Expanded polystyrene sheets
    C. Mineral wool
    D. Rubber

455. Which of the following is a variant of the cause-and-effect diagram?
    A. Histogram
    B. Pareto Chart
    C. CEDAC
    D. Flow Process Chart

Practice Questions

456. What is the most genuine method of using renewable energy for a household?
    A. Using coal energy
    B. Switching to gas energy
    C. Using domestic solar energy
    D. Relying on diesel generators

457. What does the 'S' in 5S stand for?
    A. Sort
    B. Standardize
    C. Sustain
    D. All of the above

458. What is the main disadvantage of a ground source heat pump (GSHP)?
    A. High noise level
    B. High installation cost
    C. Inefficiency in cold climates
    D. Small size

459. In the PDCA cycle, what is the purpose of the "Check" stage?
    A. To implement the plan.
    B. To compare results with targets.
    C. To formulate a plan of action.
    D. To standardize successful outcomes

460. What is a 'Green Belt' in Six Sigma?
    A. An employee with basic Six Sigma training
    B. A senior manager
    C. A financial analyst
    D. An external consultant

Practice Questions

461. Which organization has advocated for the preservation and re-purposing of buildings instead of demolition?
    A. WHO
    B. RIBA
    C. EPA
    D. UNDP

462. What should suppliers of home appliances be obliged to do?
    A. Reduce prices by 50%
    B. Produce and supply spare parts for 10 years
    C. Only sell new items without exchanges
    D. Ignore recycling efforts

463. What is the purpose of the Project Charter in the Define phase?
    A. To measure process characteristics
    B. To implement improvements
    C. To outline the project scope, objectives, and benefits
    D. To control process variations

464. What is the primary purpose of a circular economy in the context of retrofitting buildings?
    A. To increase energy consumption
    B. To eliminate waste and continually use resources
    C. To demolish old buildings
    D. To focus on non-renewable energy sources

465. Which cost is NOT part of Feigenbaum's cost of quality categories?
    A. Prevention costs
    B. Appraisal costs
    C. Internal failure costs

Practice Questions

D. Marketing costs

466. How long is the pay-back period for a typical solar energy system in the UK?
    A. 5 years
    B. 8 years
    C. 10 years
    D. 15 years

467. What does the Five S's tool focus on improving in an operation?
    A. Financial performance
    B. Employee training
    C. Housekeeping and organization
    D. Customer satisfaction

468. Six Sigma derives its name from which statistical concept?
    A. Mean deviation
    B. Standard deviation
    C. Probability distribution
    D. Variance

469. Which technique is known for using orthogonal arrays in its methodology?
    A. Failure Mode and Effects Analysis (FMEA)
    B. Design of Experiments (DOE)
    C. Quality Function Deployment (QFD)
    D. Statistical Process Control (SPC)

470. Which tool is used in the Measure phase to identify fail points in a process?
    A. Control chart

Practice Questions

B. Value stream mapping
C. Flow diagram
D. Cause and effect diagram

471. Which tool is used to visually display the frequency distribution of data?
    A. Pareto chart
    B. Scatter diagram
    C. Histogram
    D. Flow diagram

472. Which tool is recommended for detecting visual trends and identifying areas for further analysis?
    A. Run Chart
    B. Control Chart
    C. Histogram
    D. Pareto Chart

473. Which of the following is NOT a component of the DMAIC Lite methodology?
    A. Define
    B. Measure
    C. Analyze
    D. Improve and Control

474. What is a key advantage of air source heat pumps (ASHPs)?
    A. Lower installation cost
    B. Silent operation
    C. Lower energy bills
    D. Small size

Practice Questions

475. Which type of heat pump is known to be more energy-efficient and cheaper to run?
    A. Air source heat pump (ASHP)
    B. Ground source heat pump (GSHP)
    C. Water source heat pump
    D. Solar heat pump

476. Which tool is used to monitor the performance profile in the Control phase?
    A. Radar chart
    B. Cause and effect diagram
    C. Check sheet
    D. Pareto chart

477. What is the purpose of a milestone tracker diagram?
    A. To estimate project costs
    B. To show key event progress and identify slippages
    C. To optimize the critical path
    D. To calculate earned value

478. According to the Learning Objectives, what is Scrum primarily considered to be?
    A. 1 year
    B. 2 years
    C. 3 years
    D. 5 years

479. What is the primary objective of the Sustain phase in Green Six Sigma?
    A. To increase production speed
    B. To implement the solution and ensure it is sustained
    C. To reduce employee workload
    D. To improve customer satisfaction

Practice Questions

480. How much does installing an air source heat pump (ASHP) typically cost in the UK?
    A. $3,000 - $5,000
    B. $5,000 - $10,000
    C. $8,000 - $18,000
    D. $20,000 - $30,000

481. Which type of boiler is recommended for small apartments?
    A. Gas boiler
    B. Electric heat pump
    C. Stand-alone electric water heater
    D. Oil boiler

482. What does the acronym SPC stand for?
    A. Statistical Process Control
    B. Standard Process Capability
    C. Statistical Process Capability
    D. Standard Production Control

483. What is the most common material used for loft insulation?
    A. Fiberglass
    B. Mineral wool
    C. Polystyrene
    D. Rubber

484. What is the purpose of '5S' in Lean Manufacturing?
    A. To reduce production costs
    B. To organize the workplace efficiently
    C. To improve marketing strategies
    D. To train employees in statistical methods

Practice Questions

485. Which of the following is a suggested method to reduce energy consumption in homes?
    A. Use incandescent bulbs
    B. Install a smart thermostat
    C. Keep windows open during winter
    D. Use single-pane windows

486. What does the acronym TRIZ stand for?
    A. Theory of Relational Innovative Solutions
    B. Teoria Resheniya Izobreatatelskikh Zatatch
    C. Technical Risk and Innovation Strategy
    D. Test of Reliability and Innovation Systems

487. What is the final phase of the DMAICS cycle?
    A. Control
    B. Improve
    C. Sustain
    D. Verify

488. Which of the following best describes the term 'e-commerce'?
    A. Electronic exchange for buying and selling goods and services
    B. Electronic management of supply chains
    C. Electronic data interchange between employees
    D. Electronic planning of enterprise resources

489. What type of chart would you use to display how often different types of defects occur in a process?
    A. Run Chart
    B. Histogram
    C. Scatter Diagram

Practice Questions

   D. Control Chart

490. Which role would NOT typically be pursued by a Six Sigma Green Belt professional?
   A. Project Manager
   B. Marketing Director
   C. Quality Engineer
   D. Continuous Improvement Specialist

491. Which term describes a structured continuous improvement process in FIT SIGMA methodology?
   A. DMAIC Full
   B. DMAIC Lite
   C. Kaizen Event
   D. Six Sigma Event

492. Which of the following is not one of the Five S's?
   A. Seiri
   B. Seiton
   C. Seiketsu
   D. Six Sigma

493. Which of the following is NOT a dimension of service quality, according to Parasuraman et al.?
   A. Tangibles
   B. Reliability
   C. Responsiveness
   D. Innovation

494. Which tool helps in identifying and graphically displaying the root causes of a problem?

Practice Questions

    A. Run Chart
    B. Scatter Diagram
    C. Cause and Effect Diagram
    D. Control Chart

495. What percentage of heat is lost through external walls in poorly constructed homes?
    A. 10%
    B. 20%
    C. 25%
    D. 40%

496. Which of the following is NOT a deliverable of the Sustain phase?
    A. Sustain solution
    B. Sustain continuous improvement culture
    C. Sustain environment
    D. Sustain marketing strategies

497. Which country has the highest proportion of households fitted with air conditioning?
    A. US
    B. China
    C. Saudi Arabia
    D. Japan

498. What is the role of a 'Black Belt' in a Six Sigma project?
    A. Project sponsor
    B. Training coordinator
    C. Project team leader
    D. Data analyst

Practice Questions

499. Which of the following is NOT a main area of energy consumption in building operations?
    A. Heating
    B. Cooling
    C. Transportation
    D. Lighting

500. What proportion of emissions from the building sector is attributable to residential building operations?
    A. 50%
    B. 60%
    C. 72%
    D. 80%

Answers

Answers

1. Answer: B

Explanation: Every organization must have a source of income to stay in business. If an organization spends more than it takes over time, it will go out of business.

2. Answer: C

Explanation: The first formula every Six Sigma Green Belt must learn is the calculation for the S-double bar, which can be tied to the cost of quality or other financial calculations in an organization.

3. Answer: C

Explanation: DMAIC stands for Define, Measure, Analyze, Improve, and Control. It is a data-driven quality strategy used to improve processes. "Act" is not one of the phases in DMAIC; it belongs to the Plan-Do-Check-Act (PDCA) cycle, which is another process improvement methodology.

4. Answer: C

Explanation: A common reason for failure in organizational initiatives is management's lack of commitment to genuine process improvement. Without strong support and active involvement from leadership, efforts to enhance processes often falter, as they lack the necessary resources, guidance, and prioritization to succeed.

5. Answer: C

Explanation: Once the consultants leave, there are usually few internal people who understand the tools at high enough levels to continue the process effectively.

Answers

6. Answer: C

Explanation: The accounting department must be engaged early in the deployment to substantiate the cost savings being claimed or achieved.

7. Answer: C

Explanation: The video titled "Right First Time" is recommended for understanding the importance of getting things right the first time.

8. Answer: B

Explanation: At least one consultant uses the title Six Sigma Money Belt to describe a highly competent Six Sigma professional.

9. Answer: A

Explanation: Management should use Advanced Quality Planning (AQP) to prepare for a Six Sigma deployment to increase the likelihood of success within their organization.

10. Answer: B

Explanation: The group in Michigan engaged quality engineers and process engineers to improve performance in doctors' offices, which led to outstanding results.

11. Answer: A

Explanation: The Taguchi loss function measures financial loss to society resulting from poor quality, emphasizing the importance of reducing variation around the target value.

12. Answer: A

Answers

Explanation: The p-chart, developed by Walter Shewhart, was the first type of quality control chart used to monitor process behavior.

13. **Answer: B**

Explanation: Six Sigma focuses on reducing process variation and enhancing process control to deliver products and services consistently without defects.

14. **Answer: B**

Explanation: Reengineering is a breakthrough approach that involves restructuring an entire organization and its processes to improve performance.

15. **Answer: B**

Explanation: DMAIC stands for Define, Measure, Analyze, Improve, and Control, which are the steps followed in the Six Sigma problem-solving methodology.

16. **Answer: C**

Explanation: Joseph M. Juran developed the Juran trilogy, which consists of quality planning, quality control, and quality improvement.

17. **Answer: C**

Explanation: ISO 9000 is a set of international standards on quality management and quality assurance, helping companies effectively document and maintain efficient quality systems.

18. **Answer: B**

Answers

Explanation: Lean manufacturing focuses on driving out waste (non-value-added activities) and promoting work standardization and value stream mapping.

19. **Answer: D**

Explanation: Feigenbaum's three steps to quality are quality leadership, modern quality technology, and organizational commitment.

20. **Answer: B**

Explanation: The Malcolm Baldrige National Quality Award was established by the U.S. Congress in 1987 to recognize U.S. companies that have implemented successful quality management systems.

21. **Answer: C**

Explanation: Lean thinking centers on identifying and eliminating waste, thereby making clear what adds value to the customer.

22. **Answer: C**

Explanation: Toyota is credited with creating the concept of lean through their Toyota Production System (TPS) developed as far back as the 1950s.

23. **Answer: D**

Explanation: The 8Ps of lean thinking are purpose, process, people, pull, prevention, partnering, planet, and perfection. Profit is not one of them.

24. **Answer: A**

Explanation: Muda is a Japanese term used in lean thinking to describe activities that do not add value and are considered waste.

Answers

25. Answer: B

Explanation: A process step is considered value-added if it meets these three criteria: the customer recognizes the value, it transforms the product, and it is done right the first time.

26. Answer: C

Explanation: Principle 13 of The Toyota Way advises making decisions slowly by consensus, thoroughly considering all options, and then implementing decisions rapidly.

27. Answer: C

Explanation: The 5S methodology focuses on sorting, setting in order, shining, standardizing, and sustaining to improve workplace organization and efficiency.

28. Answer: A

Explanation: Jidoka, also known as autonomation, involves stopping the production process to fix problems, and ensuring that quality issues are addressed immediately.

29. Answer: D

Explanation: The TPS house components include Just-in-time, Jidoka, and Kaizen. Six Sigma is a separate methodology for process improvement.

30. Answer: B

Explanation: Takt time refers to the pace at which products must be produced to meet customer demand, aligning production with sales.

31. Answer: D

Answers

Explanation: Sustaining the efforts and continuing them require support from management and empowerment of employees. This step is often the hardest to maintain over time as it involves making 5S a part of the organizational culture.

32. Answer: B

Explanation: The Andon system is a visual feedback system that indicates the production status at any given time and empowers employees to stop the process if an issue arises that is not considered good for quality.

33. Answer: B

Explanation: Hoshin Kanri, also known as policy deployment or quality function deployment (QFD), aims to align the strategic goals of the company with the plans of middle management and the actions performed at the operational level.

34. Answer: C

Explanation: The 'Shine' step in 5S involves cleaning and inspecting the work area to ensure it is maintained in good condition and any issues are identified early.

35. Answer: B

Explanation: A Kaizen event is a focused, short-term project to improve a process and is characterized by quick, breakthrough improvements, often conducted by a cross-functional team over three to five days.

36. Answer: C

Explanation: Jidoka, also known as automation, refers to the principle of intelligent automation with a human touch, where machines can detect and

Answers

correct their operating problems, reducing the need for human intervention in repetitive tasks.

37. **Answer: B**

Explanation: One of the secrets to sustaining 5S success is to engage management by highlighting the cost savings that can be achieved through the implementation of 5S.

38. **Answer: C**

Explanation: A Kanban system is a pull system used for inventory control, ensuring that production is based on actual demand and reducing excess inventory.

39. **Answer: C**

Explanation: Gemba refers to the principle of going to the 'real place' where work is done to observe and understand operational issues firsthand, promoting deeper insights.

40. **Answer: D**

Explanation: In healthcare, a seventh 'S'—Oversight—is sometimes added to the traditional 5S methodology to ensure thorough supervision and monitoring of processes and practices.

41. **Answer: C**

Explanation: A properly administered Kanban system improves system control by assuring timely movement of products and information.

42. **Answer: D**

Answers

Explanation: Standard Work comprises standard time, standard inventory, and standard sequence. Standard cost is not a component.

43. **Answer. C**

Explanation: Takt time is used to match production with demand, setting the pace of industrial manufacturing lines according to customer demand rates.

44. **Answer: C**

Explanation: The Theory of Constraints focuses on identifying and improving the slowest process in a chain to increase the overall flow rate.

45. **Answer: B**

Explanation: The 'drum' in the Drum-Buffer-Rope analogy represents the slowest process, which sets the pace for the entire system.

46. **Answer: B**

Explanation: Poka-Yoke, or mistake-proofing, aims to prevent errors and mistakes in the production process.

47. **Answer: C**

Explanation: Overall Equipment Effectiveness (OEE) measures how effectively a manufacturing operation is utilized by multiplying availability, performance, and quality.

48. **Answer: C**

Explanation: The goal of SMED is to reduce changeover times to less than 10 minutes, thereby improving system flow and reducing production lot sizes.

Answers

49. Answer: A

Explanation: Visual Factory emphasizes the use of visual indicators to manage and display the status of material and information throughout the value stream.

50. Answer: C

Explanation: Total Productive Maintenance (TPM) ensures that equipment is reliable and helps reduce machine downtime through preventive maintenance and teamwork between maintenance technicians and line workers.

51. Answer: B

Explanation: Overproduction is defined as making more than is needed or making it earlier or faster than is needed by the next process. The principal symptom of overproduction is having excess work-in-process (WIP).

52. Answer: B

Explanation: Excess motion is caused by poor workplace layout, including awkward positioning of supplies and equipment, leading to time wasted and ergonomic issues.

53. Answer: B

Explanation: Waiting waste is typically caused by events such as delayed shipments, long setup times, or missing personnel, leading to wasted resources and potential demoralization of staff.

54. Answer: C

Answers

Explanation: Maintaining high levels of inventory incurs costs for environmental control, record keeping, and storage, which add no value to the customer.

55. **Answer: A**

Explanation: Excess movement of material/transportation is often due to poor plant layout, such as having function-oriented departments that require excessive material movement.

56. **Answer: B**

Explanation: Defect correction is non-value-added because the effort to fix defective parts does not contribute to the final product, thus wasting resources.

57. **Answer: A**

Explanation: The "sea of inventory" analogy shows that excess inventory can conceal process issues, making inefficiencies and defects less visible. Reducing inventory levels helps reveal and address these hidden problems.

58. **Answer: C**

Explanation: Cross-training personnel so they can be effectively moved to other positions helps reduce the waste of waiting by ensuring that resources are utilized efficiently.

59. **Answer: B**

Explanation: An unbalanced workload can lead to overproduction, resulting in excess work-in-process (WIP), as some parts of the process may produce more than others can handle.

Answers

60. Answer: B

Explanation: Excess processing waste involves non-value-added activities, such as dipping parts in oil to prevent rust before painting, which the customer is unwilling to pay for because it does not enhance the product.

61. Answer: B

Explanation: The goal of eliminating 'muda' is to strive for perfection by optimizing value-added activities and eliminating waste.

62. Answer: B

Explanation: Analyzing all inventory notes with an eye toward reduction or elimination is a crucial step to identify and reduce costs associated with excessive inventory.

63. Answer: B

Explanation: The grinding step following a welding operation is meant to remove weld imperfections and ensure a smooth finish.

64. Answer: B

Explanation: Lost creativity is a form of waste because it represents missed opportunities for process improvements and innovation.

65. Answer: A

Explanation: DMADV stands for Define, Measure, Analyze, Design, Verify, a methodology used in Design for Six Sigma.

66. Answer: C

Answers

Explanation: The 'Identify' phase in the IDOV process aims to link the design to customer and product requirements, ensuring alignment with the voice of the customer.

67. **Answer: D**

Explanation: The three main components of a value stream are the flow of materials, the transformation of raw materials into finished goods, and the flow of information. Customer purchase decisions are not a direct component of the value stream.

68. **Answer: B**

Explanation: Unnecessary grinding after a welding operation is classified as excessive processing, which is a form of waste in lean thinking.

69. **Answer: C**

Explanation: The 'Optimize' phase in the IDOV process develops detailed design elements to predict performance and optimize the design to minimize the sensitivity of critical-to-quality (CTQ) variables to process parameters.

70. **Answer: C**

Explanation: A value stream map uses simple graphics and icons to illustrate the movement of material, information, inventory, work-in-process, operators, and other elements within the value stream.

71. **Answer: C**

Explanation: Design for Cost focuses on minimizing production expenses by continuously seeking alternative processes, materials, and methods. This approach involves cost accounting and purchasing expertise to aid the design team in finding cost-effective solutions.

Answers

72. Answer: B

Explanation: Design for Testability requires designers to include provisions for performing tests earlier in the production cycle, rather than relying solely on functional tests of finished assemblies or subassemblies.

73. Answer: B

Explanation: MTTF stands for Mean Time to Failure, and it is used to document the expected life span of non-repairable items, indicating the average time until a failure occurs.

74. Answer: B

Explanation: Design for Scalability ensures that products or systems can handle expansion or rapid adoption without compromising quality, crucial for products deployed in growth markets.

75. Answer: C

Explanation: Design for Maintainability emphasizes the importance of making routine maintenance straightforward, which includes considerations like modularity, decoupling, and component standardization.

76. Answer: B

Explanation: The main purpose of FMEA is to systematically identify potential failure modes within a process or product, determine their effects on the system, and prioritize them based on their severity, occurrence, and detectability. This helps in identifying and addressing the most critical issues to improve reliability and prevent failures

77. Answer: C

Answers

Explanation: The Risk Priority Number (RPN) is calculated by multiplying the severity, occurrence, and detection scores, helping prioritize which failure modes require attention.

78. Answer: B

Explanation: Design for Efficiency involves creating products or systems that use minimal resources, including time, materials, and critical components, positively impacting long-term cost and reliability.

79. Answer: C

Explanation: Early involvement of manufacturing personnel helps ensure that the product design is producible and cost-effective, often through minor design changes that simplify production.

80. Answer: C

Explanation: Design for Performance aims at achieving and consistently meeting aggressive benchmarks or breakthrough performance levels, often reflected in terms like "cutting edge" or "latest and greatest."

81. Answer: B

Explanation: Six Sigma project proposals usually include preliminary measures of the seriousness of the problem, precise problem definitions, and the impact on organizational goals, but they do not typically require exact financial projections at the proposal stage.

82. Answer: C

Explanation: The performance and health of an organization are primarily gauged by its selection and use of metrics, which are often converted to financial terms such as return on investment, cost reduction, and increases in sales and/or profit.

Answers

83. Answer: B

Explanation: Advanced quality planning and quality function deployment are useful tools for setting up a project selection process in an organization that does not have one in place.

84. Answer: C

Explanation: Functional benchmarking involves comparing similar functions typically outside the organization's industry, providing opportunities to seek out benchmarking partners from different sectors.

85. Answer: A

Explanation: SIPOC stands for Suppliers, Inputs, Processes, Outputs, Customers, and is used to identify process boundaries and key elements of a process.

86. Answer: C

Explanation: Including stakeholders in process improvement teams is crucial because they possess valuable knowledge about the process, and have ideas for improvement, and their buy-in is often necessary for successful implementation.

87. Answer: B

Explanation: QFD is primarily focused on linking customer requirements to features of products and processes to meet or exceed customer expectations.

88. Answer: B

Answers

Explanation: The primary goal of benchmarking is to drive breakthrough improvement by comparing current processes with outstanding ones and setting new performance targets.

89. **Answer: B**

Explanation: After collecting customer data, it is important to verify its accuracy and consistency and resolve any conflicts or ambiguities to ensure reliable information for decision-making.

90. **Answer: C**

Explanation: Identifying the start and end points of a process is crucial for defining its boundaries, which helps improvement teams understand the scope of the process and where to focus their efforts.

91. **Answer: B**

Explanation: A negative co-relationship indicates that enhancing one technical requirement will negatively impact the other, making it worse.

92. **Answer: C**

Explanation: Column weights indicate how significant each technical requirement is in fulfilling customer requirements by multiplying the importance value with assigned relationship matrix values.

93. **Answer: D**

Explanation: In the Quality Function Deployment (QFD) matrix example, a strong relationship between customer requirements and technical descriptors is assigned a value of 9. This numerical value indicates a high level of correlation, meaning that the technical descriptor strongly supports the customer requirement.

Answers

94. Answer: C

Explanation: QFD is used to ensure that customer requirements are integrated from the initial stages of product development, enhancing customer satisfaction.

95. Answer: B

Explanation: In the context of a QFD matrix, "Whats" refers to customer requirements, which outline what the customer wants.

96. Answer: C

Explanation: When a QFD matrix has more than 25 customer voice lines, an affinity diagram can be used to condense the list and make it more manageable.

97. Answer: C

Explanation: A project charter should contain these elements to define the project's goals, limitations, and criteria for success.

98. Answer: B

Explanation: A Gantt chart is used to visualize the project timeline, showing the sequence and duration of tasks. It provides a clear overview of the project's schedule, helping to track progress and identify potential bottlenecks.

99. Answer: B

Explanation: RPN is used to prioritize project risks by assessing their potential impact (severity) and likelihood of occurrence (probability).

100. Answer: A

Answers

Explanation: In the context of Quality Function Deployment (QFD), the relationship matrix, often called the House of Quality, maps customer requirements (often referred to as "what's" or "voice of the customer") to technical requirements (often referred to as "how's" or "voice of the engineer"). This matrix helps ensure that the product development process aligns with the needs and desires of the customer by translating these needs into specific technical specifications that can be measured and met by the engineering team.

101. **Answer: B**

Explanation: Risk mitigation begins with identifying activities that can be performed to reduce the likelihood of an identified risk occurring and/or to reduce its impact if the risk is realized.

102. **Answer: B**

Explanation: Mitigation activities for a key supplier not delivering on time might include developing another supplier, providing additional supplier oversight, or sharing design information to help the supplier deliver on time.

103. **Answer: C**

Explanation: Risk mitigation plans should be reviewed regularly, usually weekly or monthly, to assess the status of existing risks and to determine if new risks have been identified or if identified risks were realized.

104. **Answer: C**

Explanation: Risk verification is the process of ensuring that the risk mitigation activities reasonably prevent the risk from occurring.

105. **Answer: B**

Answers

Explanation: If risk mitigation is not possible and the risk is realized, the team should follow the pre-planned response as identified during the risk identification, mitigation, and review processes.

106. Answer: C

Explanation: Project closure involves reviewing the project objectives achieved about the charter, ensuring that documentation is completed and stored appropriately, and conducting a closure meeting with the project sponsors.

107. Answer: A

Explanation: The project charter is an excellent tool to use as a measure of project completion as it establishes the scope, goals and objectives, and time frame for the project.

108. Answers: C

Explanation: A review of the project charter against documented project results is typically sufficient for closing a project as it ensures that all objectives and deliverables are met, stakeholder approval is obtained, necessary documentation is completed, and the project is formally closed. This comprehensive assessment confirms that the project has successfully achieved its goals and can be officially concluded.

109. Answer: B

Explanation: During project closure, it is important to identify lessons learned and inform other parts of the organization about opportunities for improvement.

110. Answer: B

Answers

Explanation: At times, the project sponsor may want an independent review or assessment before formal project closure, typically using an audit approach.

111. Answer: B

Explanation: The Activity Network Diagram (AND) is similar to the PERT chart because it graphically shows interdependencies between tasks. It highlights key tasks, the time to accomplish them, flow paths, and so on.

112. Answer: C

Explanation: Affinity diagrams are used to produce many possible answers to an open-ended question by brainstorming responses and organizing them into natural groupings.

113. Answer: B

Explanation: Advanced Quality Planning (AQP) is focused on planning to prevent surprises and save resources by considering all the elements/inputs needed to achieve the desired output.

114. Answer: B

Explanation: One of the key steps in the benchmarking process is to pilot-test ideas, ensuring that the plan will work before full implementation.

115. Answer: D

Explanation: The six Ms used in creating a cause-and-effect diagram are Man, Machine, Methods, Materials, Measurement, and Mother Nature, with Money and Management being optional.

Answers

116. Answer: B

Explanation: Auditing is an independent assessment of processes or projects against established plans and goals, often performed in organizations with quality management systems.

117. Answer: C

Explanation: An essential aspect of conducting an effective audit is being thorough in looking at the process as it relates to the standard to ensure compliance and improvement.

118. Answer: A

Explanation: Check sheets are used to collect data during process execution, which can then be analyzed to identify common patterns or trends.

119. Answer: C

Explanation: Traditional customer feedback surveys often have a low return rate, which means they do not provide a very good overall picture of customer opinions.

120. Answer: B

Explanation: The first step when starting a benchmarking project is to flowchart of the current process to ensure a clear understanding of how it works before comparing it with others.

121. Answer: B

Explanation: The first step in developing a flowchart is to define the process to be diagrammed and write its title at the top of the work surface.

Answers

122. Answer: C

Explanation: The most effective method for understanding customer needs is getting out into the world with the customers to experience what they do and to interact with them as they use your products and services.

123. Answer: C

Explanation: In a flowchart, arrows are used to show the flow of the process, indicating the sequence in which the activities occur.

124. Answer: D

Explanation: Benchmark chart is not mentioned as a type of control chart. The other options are various types of control charts used to study process changes over time.

125. Answer: A

Explanation: In an interrelationship digraph, the 'driver' is the note with the most outgoing arrows, indicating it is the most influential factor affecting the other items.

126. Answer: B

Explanation: Focus groups generally provide more accurate and in-depth responses than other techniques, although the sample is often nonrandom and too small.

127. Answer: B

Explanation: The 'Check' phase involves evaluating the results of the implemented solution to determine if the problem has been resolved satisfactorily

Answers

128. Answer: B

Explanation: After ranking each option against the criteria, the team multiplies the values by the criteria weights and calculates the row totals to determine the most favored option.

129. Answer: B

Explanation: A flowchart is a tool used to illustrate the separate steps of a process in sequential order, including inputs, outputs, decisions, and process measurements.

130. Answer: B

Explanation: The primary purpose of FMEA is to identify potential process failures and their effects to improve process reliability and quality.

131. Answer: B

Explanation: The eight-discipline (8D) approach is a systematic problem-solving method used primarily in quality management to identify, correct, and eliminate recurring problems. The first step, "Use a team approach," involves assembling a cross-functional team that has the knowledge, time, and authority to solve the problem and implement corrective actions

132. Answer: C

Explanation: PERT is used to identify and calculate the critical path for projects, which is the sequence of activities that determines the minimum project duration.

133. Answer: C

Explanation: Dr. Joseph Juran stated that 100 percent inspection is only 80 percent effective due to inspector errors and other factors.

Answers

134. Answer: B

Explanation: RPN is calculated by multiplying the severity, occurrence, and detection ratings. It helps prioritize risks based on their potential impact.

135. Answer: C

Explanation: A tree diagram breaks a general topic into several activities through a series of steps, each one more specific than the previous.

136. Answer: A

Explanation: DPU is calculated as the total number of defects divided by the total number of products produced in a given time.

137. Answer: A

Explanation: Rolled Throughput Yield (RTY) is calculated by multiplying the yields of individual processes in a series. This metric helps in understanding the cumulative effect of each process step on the overall quality, indicating the likelihood of a product passing through all stages of production without defects.

138. Answer: C

Explanation: Top-down flow is used when top management or the executive sponsor of the project is providing instructions, policies, or feedback on project performance.

139. Answer: D

Explanation: Kaoru Ishikawa emphasized respect for humanity as a management philosophy, advocating for full participatory management.

Answers

140. Answer: C

Explanation: Genichi Taguchi contributed to the philosophy of off-line quality control, focusing on designing products and processes that are robust to variations.

141. Answer: B

Explanation: The team usually comprises five to nine members, with seven being considered an ideal size. This size optimizes interaction and manageability while minimizing potential problems.

142. Answer: C

Explanation: During the Norming stage, team members resolve their conflicts and agree on mutually acceptable ideas, beginning to trust each other and share ideas and work products without hesitation.

143. Answer: C

Explanation: One of the main challenges for virtual teams is the slowing of the progression of normal team-building, which can affect commitment, buy-in, and communication, especially when nonverbal cues are missing.

144. Answer: C

Explanation: Brainstorming is a process where individuals or teams develop as many ideas as possible about a topic without criticism or judgment, focusing on quantity over quality.

145. Answer: D

Explanation: In the Performing stage, the team becomes effective with complementary skills, creating synergy and realizing their interdependence, allowing them to solve problems together and accomplish a large amount of

Answers

work.

146. Answer: B

Explanation: A Six Sigma Black Belt is responsible for leading and managing Six Sigma projects, utilizing statistical methods, and ensuring project progress by following the DMAIC process.

147. Answer: B

Explanation: During the Storming stage, the appropriate leadership style is coaching, where the leader continues close supervision and directive behavior while increasing supportive behavior and soliciting team feedback.

148. Answer: C

Explanation: The Recognition stage focuses on acknowledging and appreciating the team's achievements, which can include various forms of recognition to boost morale and motivation.

149. Answer: B

Explanation: Multi-voting is used to prioritize a list of ideas by allowing team members to vote on the most critical items, thereby narrowing down the list to focus on the most significant ideas.

150. Answer: C

Explanation: An effective countermeasure for dealing with dominant team members is to structure the agenda to provide equal participation opportunities for all team members and to moderate discussions effectively.

151. Answer: B

Answers

Explanation: Process mapping includes additional process details with the flowchart, such as costs, setup time, cycle time, inventory, and types of defects that can occur.

152. Answer: C

Explanation: Process mapping is often the first step in improving a process, as it helps visualize the entire process and identify areas for improvement.

153. Answer: C

Explanation: ISO 5807:1985 is the international standard for documentation symbols and conventions for data, program, and system flowcharts.

154. Answer: C

Explanation: In a swim lane flowchart, the process blocks are arranged in alignment with the lane of the department or function that performs a given process step.

155. Answer: C

Explanation: One common mistake in process mapping is having inadequate or inappropriate team representation for the process.

156. Answer: D

Explanation: Swim lane mapping helps visualize process steps involving multiple departments or functions by arranging process blocks in alignment with the respective lanes.

157. Answer: B

Answers

Explanation: Examples of inputs to a process include needs, ideas, expectations, and requirements. These inputs are essential for guiding the process and ensuring that the outputs align with the desired outcomes.

158. Answer: B

Explanation: A SIPOC chart provides a high-level view of a process, encompassing suppliers, inputs, processes, outputs, and customers.

159. Answer: C

Explanation: Market trends are not traditionally a category in a cause-and-effect diagram, which usually includes methods, machines, materials, etc.

160. Answer: B

Explanation: Updating the flowchart when a change is made to the process ensures compliance and accuracy, and helps in maintaining the process's integrity and performance.

161. Answer: B

Explanation: Brainstorming intentionally encourages divergent thinking through which most possible causes are identified, making it a powerful technique for soliciting ideas.

162. Answer: C

Explanation: NGT is effective when the issue is sensitive and emotional or when the team is composed of members from several layers of the organization, encouraging all team members to contribute without pressure.

163. Answer: C

Answers

Explanation: The Pareto chart helps visualize the items charted as "vital few" and "trivial many," adhering to the 80:20 principle which is crucial for focusing on the most significant issues.

164. Answer: C

Explanation: CEDAC involves displaying the fishbone diagram on a wall or board and encouraging participants to write causes on it or use sticky notes, depending on organizational culture and communication.

165. Answer: B

Explanation: The affinity diagram helps organize a large number of ideas into their natural relationships, making it easier for the team to grasp complex issues and achieve group consensus.

166. Answer: B

Explanation: A true Pareto chart includes data arranged in descending order of frequency, a secondary axis with percentages, and a cumulative percentage line, making it more informative and useful.

167. Answer: C

Explanation: Criticizing ideas instantly and discarding them can discourage team members from contributing, which is counterproductive to effective idea generation in brainstorming.

168. Answer: A

Explanation: The facilitator's job in brainstorming is to enforce ground rules and encourage ideas, ensuring a free flow of thoughts without hesitation or judgment.

Answers

169. Answer: A

Explanation: The relationship diagram is used to display the degree of the relationship between variables, causes, and effects, which helps in prioritizing and understanding the impact of different factors.

170. Answer: C

Explanation: To perform an effective Pareto analysis, the data should not be too specific, as this can result in a poor Pareto chart with many trivial items, defeating the purpose of identifying the "vital few" issues.

171. Answer: C

Explanation: Statisticians usually consider a sample size of 30 or more to be sufficiently large for the sampling distribution of the mean to approximate a normal distribution, as per the central limit theorem.

172. Answer: C

Explanation:

The standard error of the mean (SEM) quantifies the variability of the sample mean estimates from different samples drawn from the same population. It is calculated as the standard deviation of the population divided by the square root of the sample size. The SEM gives an idea of how much the sample mean would vary if you were to take multiple samples from the population.

173. Answer: C

Explanation: The central limit theorem states that the sampling distribution of the mean will increasingly approximate a normal distribution as the sample size increases, regardless of the population's distribution.

Answers

174. Answer: B

Explanation: In a normal distribution, approximately 95% of the data falls within ±1.96 standard deviations from the mean, which is a common interval used for confidence intervals.

175. Answer: A

Explanation: The 95% confidence interval is calculated as the sample mean ± 1.96 times the standard error: 10 ± 1.96 * 0.003 = 9.9941 to 10.0059 milligrams.

176. Answer: B

Explanation: Inferential statistics use data from a sample to make estimates or inferences about the population from which the sample was drawn.

177. Answer: B

Explanation: Control charts use the central limit theorem to monitor the status of a process by averaging the measurement values in subgroups, making them robust to departures from normality.

178. Answer: B

Explanation: The sample standard deviation is calculated as the population standard deviation divided by the square root of the sample size: $0.012 / \sqrt{16}$ = 0.003 milligrams.

179. Answer: B

Explanation: The calculation uses the binomial formula to find the probability (P) of getting exactly 1 success (x = 1) in 5 trials (n = 5) where the chance of success in each trial is 10% (p = 0.1). The result, 0.328, represents the likelihood of this specific scenario.

Answers

180. Answer: C

Explanation: The Poisson distribution is used to model the number of defects on a painted surface and other similar discrete data.

181. Answer: D

Explanation: Continuous measurements provide more detailed and precise information about variations in a process over time compared to discrete measurements, which are more limited in their granularity. This sensitivity to process changes allows for finer control and understanding of variability, which is crucial in quality management and process improvement efforts.

182. Answer: C

Explanation: Continuous scales are characterized by the infinite number of possible values between any two measurements, allowing for high precision.

183. Answer: C

Explanation: Surveys often have low response rates, approximately 10 to 15 percent, which can affect the reliability of the collected data.

184. Answer: C

Explanation: Stratified sampling is used when the lot is heterogeneous, by dividing it into subgroups (stratA) and then randomly sampling from each group.

185. Answer: C

Explanation: Real-time data acquisition minimizes human errors and allows for prompt decision-making, reducing the impact on production or service quality.

Answers

186. Answer: B

Explanation: Locational data identify where data are coming from, such as pinpointing paint defects in specific areas on an automobile.

187. Answer: B

Explanation: Multiple points of data entry increase the risk of inconsistency and errors, making it crucial to have a streamlined data collection process.

188. Answer: C

Explanation: Ratio scales have a true zero point, allowing for multiplication and division operations, making them suitable for more complex statistical analyses.

189. Answer: B

Explanation: Repeatability and reproducibility (R&R) studies help ensure the accuracy and consistency of the measurement system, reducing errors in data entry.

190. Answer: B

Explanation: Check sheets are designed to pre-categorize potential outcomes, facilitating easy and consistent data collection during process execution.

191. Answer: B

Explanation: A check sheet helps focus on continual improvement and may foster changes just because it is being used. It is also the basis for other analytical tools and is incorporated into attribute statistical process control charts.

Answers

192. Answer: D

Explanation: The range is a measure of dispersion, not central tendency. Measures of central tendency include the mean, median, and mode.

193. Answer: D

Explanation: The sample standard deviation produces an estimate of the standard deviation of the population from which the sample was drawn.

194. Answer: A

Explanation: The upper whisker extends to the highest data value within the upper limit, which is calculated as Q3 + 1.5(IQR).

195. Answer: C

Explanation: The Weibull distribution is used for reliability data when the underlying distribution is unknown. It helps estimate the shape parameter b and MTBF.

196. Answer: C

Explanation: A scatter diagram is a powerful visual tool used to display relationships or associations between two variables, cause and effect, etc.

197. Answer: C

Explanation: The Anderson-Darling test is most widely used by statistical software to check the normality of random data.

198. Answer: C

Answers

Explanation: In a stem-and-leaf plot, the count for the row containing the median value is enclosed in parentheses.

199. Answer: B

Explanation: A run chart is used to identify patterns in-process data and detect any nonrandom behavior such as trends, oscillation, mixtures, and clustering.

200. Answer: D

Explanation: A box-and-whisker plot, also known as a box plot, uses the high and low values of the data as well as the quartiles to provide a pictorial view of the data distribution.

201. Answer: B

Explanation: The primary purpose of MSA is to analyze and reduce the variation in the measurement system, ensuring the accuracy and reliability of measurement data.

202. Answer: B

Explanation: Repeatability refers to the variation in measurement when measured by one appraiser on the same equipment in the same measurement setting at the same time.

203. Answer: C

Explanation: The design of the measurement system influences repeatability, not reproducibility. Reproducibility is influenced by operator training, skill, knowledge, and consistency in measurement.

204. Answer: C

Answers

Explanation: The first step in conducting a GR&R study is to plan the study in detail, ensuring that all necessary resources and conditions are met.

205. Answer: B

Explanation: In MSA, bias refers to the difference between an absolute value and the true value concerning a standard master at various measurement points.

206. Answer: A

Explanation: Ensuring that appraisers measure all samples for every trial is crucial for the completeness of the study and the accuracy of the calculations.

207. Answer: B

Explanation: Precision in MSA refers to the closeness of repeated readings to each other, representing the consistency of the measurement system.

208. Answer: D

Explanation: Linearity refers to the accuracy of measurement at various measurement points of the measuring range in the equipment.

209. Answer: B

Explanation: If points in the sample range charts are outside the control limits, it is important to investigate whether this is due to a recording error or any special causes.

210. Answer: A

Answers

Explanation: In the Gage Repeatability and Reproducibility Report, 'EV' stands for Equipment Variation, which is an estimate of the standard deviation of the variation due to repeatability.

211. Answer: C

Explanation: The AIAG MSA Manual defines discrimination as the measurement resolution, scale limit, or smallest detectable unit of the measurement device and standard.

212. Answer: C

Explanation: A number of distinct categories greater than or equal to 5 is considered acceptable for process monitoring applications.

213. Answer: B

Explanation: Reproducibility refers to the variation in measurements when different appraisers measure the same part using the same measurement system. It assesses the consistency of the measurement system across different operators.

214. Answer: C

Explanation: The higher the gage R&R (Gage Repeatability and Reproducibility), the higher the measurement error, which leads to a higher error in the Cp (Capability Index) assessment. This is because increased variability in measurements reduces the accuracy and reliability of the process capability assessment.

215. Answer: A

Explanation: Not randomizing the samples during measurement is a common error, whereas randomizing them is recommended.

Answers

216. Answer: C

Explanation: Randomizing the samples during measurement helps avoid introducing bias in repetitive measurement trials.

217. Answer: B

Explanation: It is recommended to present the results anonymously to avoid unhealthy comparisons between appraisers.

218. Answer: B

Explanation: Nonlinearity can be due to improper calibration at the high and low ends of the operating range, so proper calibration is necessary.

219. Answer: C

Explanation: GR&R (Gage Repeatability and Reproducibility) results need to be periodically validated to ensure ongoing measurement accuracy and reliability, similar to the regular calibration of gages. This helps maintain the integrity of the measurement system over time.

220. Answer: C

Explanation: Scatter diagrams and correlation coefficients are used to compare multiple measurement devices and check for consistency.

221. Answer: C

Explanation: The key point of ongoing measurement system analysis is to understand the uncertainty of the measurement system, which helps in understanding what exactly is being measured.

222. Answer: A

Answers

Explanation: In orthogonal regression, if the intercept is close to 0 and the slope is close to 1, it suggests that the two methods being compared provide similar measurements, indicating equivalence.

223. Answer: C

Explanation: Appraiser C showed the highest within-appraiser assessment agreement with a percent agreement of 84.4%, as per the provided data.

224. Answer: D

Explanation: Both Appraisers B and C had the highest agreement with the standard, each with a percent agreement of 78.1%.

225. Answer: C

Explanation: A significant p-value in the variance between measuring equipment indicates that there might be an issue that needs further investigation by the experimenter.

226. Answer: B

Explanation: GR&R stands for Gauge Repeatability & Reproducibility, which is a method used to assess the precision of measurement systems.

227. Answer: D

Explanation: The high cost of raw materials is not listed as a practical challenge in GR&R studies. Challenges include one-sided specification, skewed distribution, and fully automated equipment.

228. Answer: A

Explanation: The 95% confidence interval indicates the range within which the true value lies with 95% certainty, providing a measure of the reliability of the estimate.

Answers

229. Answer: C

Explanation: Appraiser C is more consistent in judgment but also accepted 33.3% of bad parts, which is higher than the other appraisers.

230. Answer: B

Explanation: If the process is not in statistical control and does not meet specification, the appropriate action is to stop the process and immediately investigate to identify and correct the issue.

231. Answer: B

Explanation: The first step in a process capability study is to perform measurement system analysis (MSA). This ensures that the measurement system variation does not mislead the capability assessment and stability monitoring.

232. Answer: C

Explanation: Assignable (special) causes are specific, identifiable sources of variation, such as significant changes in incoming material quality, that are not part of the inherent process variation.

233. Answer: B

Explanation: Organizations may create internal specifications tighter than customer specifications to ensure higher quality and provide additional protection to customers.

234. Answer: B

Explanation: ZU represents the number of standard deviations between the process average (X) and the upper specification limit (USL).

Answers

235. Answer: B

Explanation: Control charts are used to monitor the stability and consistency of a process over time. Points that fall outside the control limits, typically set at ±3 standard deviations from the centerline (mean), suggest that the process may be experiencing special cause variation or non-random variation. This signifies that something unusual or unexpected has occurred in the process, indicating that corrective action may be needed to investigate and address the issue causing the out-of-control condition.

236. Answer: B

Explanation: Ensuring the range chart is within control is crucial because it confirms that the variation is stable before any adjustments are made to the process averages.

237. Answer: B

Explanation: The natural process limits are calculated as ±3 standard deviations from the process average. For this example, it is $1.065 \pm 3(0.024)$ = 0.993 to 1.137 lbs.

238. Answer: B

Explanation: Engineers may be hesitant to invest time in investigating out-of-control conditions because it requires a commitment of time and resources.

239. Answer: B

Explanation: Test 4 indicates a pattern where 14 points in a row alternate up and down, which is a signal of an out-of-control condition.

240. Answer: D

Answers

Explanation: Assignable (special) causes should be investigated and removed before estimating process capability to ensure the process is stable.

241. Answer: B

Explanation: Multi-vari studies are used to understand variations within a process both analytically and graphically. They help identify and minimize excessive variations by analyzing cyclical, temporal, and positional variations.

242. Answer: D

Explanation: The three types of variation analyzed in multi-vari studies are cyclical (part-to-part or lot-to-lot), temporal (shift-to-shift), and positional (within-part).

243. Answer: A

Explanation: Positional variation refers to the variation of a characteristic on the same product and is also known as within-part variation.

244. Answer: C

Explanation: Temporal variation, also known as shift-to-shift variation, occurs as changes in the measurement process over different periods or shifts. This type of variation can impact the consistency and reliability of measurements.

245. Answer: B

Explanation: The multi-vari sampling plan procedure includes recording the time and values from each sample set in a table format, selecting a manageable sample size, plotting a graph with time on the horizontal scale, and measuring values on the vertical scale.

Answers

246. Answer: C

Explanation: The initial multi-vari study revealed that the most significant variation was caused by part-to-part differences, which led the team to investigate the casting process for improvements.

247. Answer: B

Explanation: The new metal foundry was sourced to perform precision casting, which aimed to reduce part-to-part variation in the stainless steel casting process.

248. Answer: D

Explanation: The team suggested removing the parts from the lathe and pressure-washing them before the burnishing step to prevent machining chips from being burnished into the surface.

249. Answer: B

Explanation: The remaining shift-to-shift variation was suspected to be due to tool wear, leading to the implementation of a tooling wear verification and change process.

250. Answer: B

Explanation: In a transactional process, data should be stratified by transaction type and complexity before applying a multi-vari study to understand the variation in a meaningful way.

251. Answer: B

Explanation: When a cause-and-effect relationship exists (causation), there will always be a correlation between the variables. However, correlation alone does not imply causation.

Answers

252. Answer: C

Explanation: The correlation coefficient, r, ranges from -1.0 to +1.0, indicating the strength and direction of the relationship between the two variables.

253. Answer: C

Explanation: A correlation coefficient close to zero suggests that there is no linear relationship between the two variables.

254. Answer: B

Explanation: The convention is to place the independent (x) variable on the horizontal axis and the dependent (y) variable on the vertical axis in a scatter plot.

255. Answer: B

Explanation: A p-value less than 0.05 indicates that there is strong evidence against the null hypothesis, leading to its rejection.

256. Answer: D

Explanation: An r value of +1.0 indicates a perfect positive linear relationship between the two variables, meaning as one variable increases, the other variable increases proportionally.

257. Answer: C

Explanation: The equation for a linear regression line is (y = a + bx), where (a) is the y-intercept and (b) is the slope. This equation represents the relationship between the dependent variable (y) and the independent variable (x).

Answers

258. Answer: C

Explanation: The slope (B) represents the change in the dependent variable (y) for each one-unit change in the independent variable (x).

259. Answer: B

Explanation: The standard error of the estimate (se) measures how much the observed values deviate from the values predicted by the regression line.

260. Answer: C

Explanation: The CORREL function in Excel is used to calculate the correlation coefficient between two sets of data.

261. Answer: C

Explanation: Including more than one independent variable allows the model to explain a higher proportion of the variation in the dependent variable.

262. Answer: A

Explanation: The equation $y = b_0 + b_1 x$ represents the linear relationship where b_0 is the y-intercept and b_1 is the slope of the line.

263. Answer: A

Explanation: The null hypothesis represents the status quo and is believed to be true unless there is overwhelming evidence to the contrary.

264. Answer: A

Answers

Explanation: A Type I error occurs when the null hypothesis is rejected when it is actually true. This means that we incorrectly conclude there is an effect or difference when in fact, there is not one.

265. Answer: C

Explanation: If the test statistic falls within the rejection region, there is enough evidence to reject the null hypothesis.

266. Answer: B

Explanation: The p-value indicates the probability of obtaining a test statistic as extreme as, or more extreme than, the one observed, assuming the null hypothesis is true.

267. Answer: B

Explanation: For a one-sample Z-test, the test statistic is calculated as $Z = (X - m_o) / (s / \sqrt{n})$ where X is the sample mean, m_o is the population mean, s is the standard deviation, and n is the sample size.

268. Answer: A

Explanation: For large samples, the confidence interval for the mean is calculated as $X \pm Z (s / \sqrt{n})$, where X is the sample mean, Z is the Z-value for the desired confidence level, s is the population standard deviation, and n is the sample size.

269. Answer: B

Explanation: When the population variance is unknown and the sample size is small, the test statistic for a t-test is calculated as $t = (X - m_o) / (s / \sqrt{n})$.

Copyright © 2024 VERSAtile Reads. All rights reserved.
This material is protected by copyright, any infringement will be dealt with legal and punitive action.

Answers

270. Answer: C

Explanation: The t-distribution is used for samples drawn from a normally distributed population when the population standard deviation is unknown. It depends on the degrees of freedom, which are related to the sample size.

271. Answer: B

Explanation: The null hypothesis for the chi-square test for variance provided in the context is σ-squared = 36, which is stated directly in the problem setup.

272. Answer: B

Explanation: The p-value for the chi-square test for variance being given as 0.257 in the Minitab analysis indicates that there is no statistically significant difference in variance at common significance levels (e.g., 0.05 or 0.01). This means that the null hypothesis, which typically states that the variance is equal to a specified value, cannot be rejected based on the data.

273. Answer: C

Explanation: If the p-value is greater than the significance level (0.05), we fail to reject the null hypothesis. Therefore, we conclude there is no significant difference in variances.

274. Answer: A

Explanation: The 95% confidence interval for the variance, as given in the Minitab analysis, being (30.1, 79.0) means that we are 95% confident that the true population variance lies within this range. This interval provides a range of plausible values for the variance based on the sample data.

275. Answer: D

Answers

Explanation: The chi-square distribution is used to determine the confidence interval for the variance, as indicated in the context.

276. Answer: A

Explanation. The sample variance for the set of 35 samples being given as 46 means that the average of the squared deviations from the mean for these 35 samples is 46. This value quantifies the degree of spread or dispersion in the sample data.

277. Answer: B

Explanation: The chi-square test for variance requires that the population from which the samples are drawn must be normally distributed.

278. Answer: B

Explanation: The chi-square statistic value for the variance test with 34 degrees of freedom is 43.44 as per the Minitab Analysis.

279. Answer: B

Explanation: Since the variance falls within the interval (30.1, 79.0) and the p-value is greater than 0.05, the analysis concludes there is no significant difference.

280. Answer: A

Explanation: The standard deviation range at a 95% confidence interval according to the Minitab Analysis is (5.49, 8.89).

281. Answer: B

Answers

Explanation: Levels refer to the settings or possible values of a factor in an experimental design. These can be quantitative measures (e.g., three different temperatures) or qualitative (e.g., high–medium–low).

282. Answer: C

Explanation: Replication repeats observations or measurements to increase precision, reduce measurement errors, and balance unknown factors. This involves repeating the set of all treatment combinations in an experiment.

283. Answer: C

Explanation: Dependent variables are variables that depend on another variable. An independent variable is said to affect or cause changes in the dependent variable.

284. Answer: C

Explanation: Randomization organizes the experiment so that treatment combinations are done in a chance manner, improving statistical validity.

285. Answer: A

Explanation: A block is a portion of the experimental material or environment common to itself and distinct from other portions, such as samples from the same batch.

286. Answer: B

Explanation: Replication is the process of running experimental trials in a random manner, while repetition involves running trials under the same setup of machine parameters.

287. Answer: B

Answers

Explanation: Experimental error is the variation in the response variable when levels and factors are held constant. It must be subtracted to determine the true effect of an experiment.

288. Answer: C

Explanation: A treatment combination is the set or series of levels for all factors in a given experimental run, e.g., pressure—200 psi, temperature—70 °F, feed—high.

289. Answer: B

Explanation: The primary objective of a designed experiment is to generate knowledge about a product or process, specifically to find the effect that a set of independent variables has on a set of dependent variables.

290. Answer: B

Explanation: Response variables are variables that show the observed results of an experimental treatment. The response is the outcome of the experiment as a result of controlling the levels, interactions, and number of factors.

291. Answer: B

Explanation: A completely randomized experiment is appropriate when only one factor is analyzed, allowing for clear separation of that factor's effects.

292. Answer: C

Explanation: Factorial designs are appropriate when investigating several factors at two or more levels and when the interactions between these factors are necessary to understand.

Answers

293. Answer: C

Explanation: A response variable refers to the variables influenced by the other factors, for which experimental objectives are set.

294. Answer: B

Explanation: Randomization is a method used to lessen the effects of special cause variation by assigning the order of tests randomly.

295. Answer: C

Explanation: Partially balanced incomplete blocks expand on balanced incomplete blocks when a larger number of blocks are required for the experiment.

296. Answer: C

Explanation: The first consideration is understanding the question or objective that the experiment seeks to answer, as this guides the entire experimental design process.

297. Answer: B

Explanation: In a Latin square design, the number of rows, columns, and treatments must all be equal, with no interactions between them.

298. Answer: B

Explanation: Mixture designs are factorial experiments where factor levels are expressed in percentages that add up to 100%, focusing on the proportions of mixture components.

Answers

299. Answer: B

Explanation: Nested designs are suitable for studying relative variability instead of mean effects, providing a way to separate different sources of variability.

300. Answer: C

Explanation: A Latin square design is used to control the interactions and effects of other variables while focusing on a primary factor, ensuring no interaction between rows, columns, and the studied factor.

301. Answer: B

Explanation: The primary purpose of running a DOE is to understand the variation in a process through testing and optimizing, thereby determining better ways of doing things or understanding other factors in the process.

302. Answer: C

Explanation: Blocking is a method of designing and organizing the experiment that attempts to lessen the effects of special cause variation by grouping the experiments in batches of tests or runs.

303. Answer: B

Explanation: Interaction occurs when one or more factors influence the behavior of another factor.

304. Answer: B

Explanation: You can nullify the impact of a shift change by doing the first three replicates of each run during the first shift and the remaining two replicates of each run during the second shift.

Answers

305. Answer: B

Explanation: An experimental design is called balanced when each setting of each factor appears the same number of times with each setting of every other factor.

306. Answer: B

Explanation: Confirmation tests are conducted to determine if the experiment has successfully identified better process conditions.

307. Answer: C

Explanation: Confounding occurs when a factor interaction cannot be separately determined from a major factor in an experiment.

308. Answer: D

Explanation: Brainstorming the key factors causing variation is a crucial step in DOE to identify and categorize sources of variation.

309. Answer: B

Explanation: In a completely randomized design, experimental tests are randomly assigned to subjects or experimental units without any systematic bias. This randomization ensures that each test or treatment has an equal chance of being assigned to any subject, reducing the potential for bias and allowing for valid statistical inference based on the results.

310. Answer: B

Explanation: ANOVA procedures are used to determine whether the difference between high and low values are due to a real difference in the dependent variable or due to experimental error.

Answers

311. Answer: C

Explanation: In a full factorial experiment, the number of runs is determined by the formula (L^F), where L is the number of levels and F is the number of factors. This ensures that every possible combination of factors and levels is tested.

312. Answer: A

Explanation: The effect of the cooking method was calculated to be 23.25, which is much higher than the effects of meat (-5.75) and marinade (2.25). This indicates that the cooking method had the greatest impact on the approval rating.

313. Answer: D

Explanation: The Fishbone Diagram, also known as the cause-and-effect or Ishikawa diagram, is widely used for identifying the root causes of a problem by categorizing potential causes systematically.

314. Answer: C

Explanation: A SIPOC diagram (Suppliers, Inputs, Process, Outputs, Customers) is used to provide a high-level overview of a process. It helps to define the process boundaries, identify key elements, and clarify the scope before detailed analysis. This tool is often used during the Define phase of DMAIC

315. Answer: B

Explanation: Fractional factorial experiments are designed to save time and resources by not examining every possible combination, making them suitable for quick exploratory tests where interactions are insignificant.

316. Answer: B

Answers

Explanation: ANOVA is used to assess the significance of the experimental effects in a two-level fractional factorial experiment. It helps determine whether changing the levels of factors significantly affects the process outcomes.

317. Answer: C

Explanation: Root cause analysis focuses on implementing permanent solutions rather than temporary ones. The phases include defining the problem, collecting and analyzing data, and establishing a corrective action plan.

318. Answer: D

Explanation: The phase "Standardize the Solution" involves making necessary modifications to control plans, process references, procedures, and acceptance criteria to ensure the solution is maintained and prevents recurrence.

319. Answer: C

Explanation: The 5 Whys method involves asking "why" multiple times (typically five) to drill down to the root cause of a problem. It helps uncover deeper layers of a problem that might not be immediately apparent.

320. Answer: C

Explanation: A cause-and-effect (X-Y) relational matrix is used to quantify and prioritize the impacts of causes (X) on effects (Y) through numerical ranking, helping to identify which causes have the greatest impact and should be addressed first.

321. Answer: C

Explanation: The 5S methodology consists of Sort, Straighten, Shine, Standardize, and Sustain. Simplify is not one of the 5S steps.

Answers

322. Answer: B

Explanation: Pull systems manage scheduling and material flow to eliminate overproduction and excess inventories based on actual customer demand.

323. Answer: B

Explanation: Poka-yoke refers to mistake-proofing methods designed to prevent errors from occurring in a process.

324. Answer: B

Explanation: Lean reduces the quantity of work-in-process (WIP) inventory, which reduces the need for large, segregated inventory areas, thus lowering carrying costs.

325. Answer: B

Explanation: 5S stands for sort, straighten, shine, standardize, and sustain, and is used to create order and organization in the workspace.

326. Answer: B

Explanation: TPM aims to manage operations with preventive maintenance, load balancing, and streamlined flow control to avoid performance reductions and bottlenecks.

327. Answer: A

Explanation: The Visual Factory applies visual displays and controls to allow anyone to promptly know the status of the process and interpret whether it is operating properly.

Answers

328. Answer: B

Explanation: Kaizen is focused on continuous improvement, often through low-cost and incremental changes to processes.

329. Answer: B

Explanation: Standard work provides a baseline for performing tasks and activities, helping to identify and demonstrate improvements in quality, cycle time, and waste reduction.

330. Answer: B

Explanation: Process capability is a measure of how well a process can meet customer needs by comparing VOC with VOP. The higher the result, the more capable the process is of satisfying customer requirements.

331. Answer: B

Explanation: A process capacity sheet helps in calculating and understanding the capacity of equipment by considering inputs like machine time per piece, manual time per piece, and tool change time per piece.

332. Answer: C

Explanation: Workflow analysis evaluates how well processes within a system are linked and work together to achieve an expected outcome, focusing on improving efficiency and effectiveness.

333. Answer: C

Explanation: Takt time helps in matching the pace of production to customer demand, ensuring that the process meets customer requirements without creating backlogs.

Answers

334. Answer: C

Explanation: One-piece flow aims to move one work piece at a time between operations within a work cell, ensuring quality and minimizing queue times and interruptions.

335. Answer: C

Explanation: The TOC approach identifies bottlenecks in the process and focuses on balancing workflows to improve overall efficiency and reduce delays.

336. Answer: C

Explanation: Value stream mapping illustrates the sequential activities within a value stream, showing where value is created and where waste exists, enabling the identification of value-added and non-value-added activities.

337. Answer: B

Explanation: Quick changeover/setup reduction aims to minimize the time required for changing over production equipment, thereby increasing flexibility and reducing bottlenecks.

338. Answer: B

Explanation: Kanban is a replenishment system that orders new material only after the existing material is consumed, helping to manage inventory efficiently.

339. Answer: B

Answers

Explanation: Kaizen is a Japanese term for continuous improvement. Its primary objective is to promote small, easy, low-risk, and low-cost process improvements that can be quickly implemented.

340. Answer: C

Explanation: A Kaizen Blitz accelerates the pace of Kaizen by completing a continuous improvement project within a week through workshops and cross-functional team efforts.

341. Answer: C

Explanation: A CCR is not the same as a bottleneck. While a bottleneck is a specific point of congestion that limits throughput, a CCR impacts processes even when they are not running at full capacity.

342. Answer: C

Explanation: Gemba Kaizen focuses on making improvements at "Gemba," the real place where work happens, involving those who know the process best.

343. Answer: C

Explanation: Kaikaku refers to major redesigns of processes, products, or business strategies, making it revolutionary as opposed to the evolutionary nature of Kaizen.

344. Answer: A

Explanation: Rational subgrouping in control charts involves grouping data points that are collected under similar conditions or circumstances. The purpose is to ensure that the variation observed within each subgroup is primarily due to common causes (random variation) rather than special causes (non-random variation). This approach helps in accurately assessing process stability and performance over time, facilitating effective decision-

Answers

making and improvement efforts based on the data trends observed.

345. Answer: B

Explanation: The primary goal of SPC is to monitor and improve process performance by measuring and analyzing variations in processes

346. Answer: B

Explanation: A state of statistical control means that only random causes of variation are present in the process, not that the products necessarily meet specifications.

347. Answer: D

Explanation: Follow-through is essential for ensuring that improvements are maintained and preventing backsliding. Ignoring follow-through would hinder Kaizen success.

348. Answer: B

Explanation: "Gemba" is a Japanese term that means "the real place" where work happens, emphasizing the importance of making improvements at the actual site of operations.

349. Answer: C

Explanation: Under-adjustment or under-control occurs when a process fails to respond to the presence of a special cause of variation. This can lead to additional process variation.

350. Answer: C

Explanation: Continuous data are used in control charts for variables, which incorporate measurements of central tendency (averages) or variation (range).

Answers

351.   **Answer:** C

Explanation: The main goal of Six Sigma is to enhance business processes by minimizing defects and errors, reducing variation, and increasing quality and efficiency, aiming for only 3.4 defects per million opportunities.

352.   **Answer:** D

Explanation: A histogram is a single tool used to display the distribution of data and is often used in quality management processes.

353.   **Answer:** C

Explanation: The Define phase involves forming a project team and drafting a project charter, which includes the definition and scope of the problem, project objectives, and benefits.

354.   **Answer:** B

Explanation: Driving forces support change, whereas restraining forces resist change.

355.   **Answer:** C

Explanation: The Process Capability Index (Cp) measures the inherent reproducibility of a process.

356.   **Answer:** B

Explanation: While it is not mandatory to undergo Lean Six Sigma training before taking the exam, it is recommended.

Answers

357.   Answer: B

Explanation: The Interrelationship Diagram emphasizes thinking in multiple directions to analyze and classify cause-and-effect relationships among critical issues.

358.   Answer: C

Explanation: Six Sigma aims to improve business processes by identifying and removing the causes of defects and minimizing variability.

359.   Answer: B

Explanation: The Improve phase is focused on developing solutions to improve process performance and validating that these solutions meet or exceed quality improvement goals.

360.   Answer: B

Explanation: The Control phase is about maintaining the improvements achieved during the Improve phase.

361.   Answer: C

Explanation: Green Six Sigma aims to achieve sustainable results by integrating quality management and improvement programs with a focus on environmental sustainability.

362.   Answer: B

Explanation: The primary goal of DFSS is to design products and processes that meet or exceed customer needs and CTQ requirements from the outset.

Answers

363.   Answer: B

Explanation: DMAIC is a structured approach used in Six Sigma for process improvement, which stands for Define, Measure, Analyze, Improve, and Control.

364.   Answer: B

Explanation: In the Measure phase, a detailed flow diagram is used to identify fail points in a process.

365.   Answer: C

Explanation: The barriers mentioned include a packaged approach, high start-up costs, and top-down directives, but not frequent changes in technology.

366.   Answer: B

Explanation: The affinity diagram is also known as the KJ method, identified with Kawakita Jiro, the Japanese scientist who first applied it in the 1950s.

367.   Answer: C

Explanation: OEE stands for Overall Equipment Effectiveness, which measures the delivered performance of plant or equipment based on good output.

368.   Answer: B

Explanation: The IASSC provides the IASSC Certified Lean Six Sigma Green Belt™ (ICGB™) certification for Green Belts.

Answers

369.   Answer: B

Explanation: A Gantt chart is also known as a bar chart.

370.   Answer: C

Explanation: Points outside the control limits indicate special cause variation that is not a part of the normal process.

371.   Answer: B

Explanation: Depending on the organization, Lean Six Sigma Green Belts may spend between 10-25% of their time on Lean Six Sigma projects.

372.   Answer: B

Explanation: The Define phase is crucial for setting the groundwork of a Green Six Sigma project by clearly identifying and prioritizing the right project to focus on.

373.   Answer: C

Explanation: FIT SIGMA is synonymous with 'FIT' and is a solution for sustainable excellence in all operations.

374.   Answer: D

Explanation: Blockchain technology ensures transparency and traceability, making it known as 'trustware'.

375.   Answer: B

Answers

Explanation: Process capability refers to the ability of a process to produce output that meets specifications.

376. **Answer**: A

Explanation: DMAIC stands for Define, Measure, Analyze, Improve, Control, which is the life cycle approach for Six Sigma projects.

377. **Answer**: D

Explanation: Flow diagrams are typically used in the Define phase, while control charts, scatter diagrams, and check sheets are used in the Measure phase.

378. **Answer**: B

Explanation: Mistake proofing, or poka-yoke, aims to prevent errors and detect defects to improve overall quality.

379. **Answer**: B

Explanation: A PESTLE analysis focuses on assessing the impact of external contexts, including political, economic, social, technological, legal, and environmental factors, on a project or major operation.

380. **Answer**: C

Explanation: The phases outlined by the IASSC Lean Six Sigma Green Belt Body of Knowledge include Define, Measure, Analyze, Improve, and Control. "Execute" is not one of the phases.

381. **Answer**: C

Answers

Explanation: EVM stands for Earned Value Management. Earned Value Management (EVM) is a project management methodology used to measure project performance and progress. It integrates project scope, schedule, and cost variables to provide accurate forecasts and identify project variances. EVM helps project managers understand the true status of a project and make informed decisions to keep it on track.

382. **Answer:** B

Explanation: These cornerstones form the fundamentals of the FIT methodology.

383. **Answer:** C

Explanation: Non-renewable energy is not listed as an urgently required improved technology for climate change initiatives; the focus is on zero-carbon and renewable solutions.

384. **Answer:** B

Explanation: Scatter diagrams are used to study and identify possible relationships between two variables.

385. **Answer:** B

Explanation: RPN is the product of the severity, occurrence, and detection ratings used to prioritize potential failure modes.

386. **Answer:** B

Explanation: The IASSC Certified Lean Six Sigma Green Belt Exam™ consists of 100 questions, with an additional 10 non-graded questions in certain versions.

Answers

387.   Answer: C

Explanation: The UK Government announced a $6.2 billion program to protect properties from flooding by 2027.

388.   Answer: A

Explanation: Bill Gates suggests that CGIAR can help farmers adapt to climate change with a wider variety of crops and livestock.

389.   Answer: B

Explanation: The activity network diagram helps identify the critical path, highlighting the most critical tasks that might delay a project.

390.   Answer: D

Explanation: The Control phase follows the Improve phase and focuses on maintaining the improvements made.

391.   Answer: B

Explanation: The reduction in fatalities during Cyclone Amphan in Bangladesh is attributed to early warning systems and concrete cyclone shelters.

392.   Answer: C

Explanation: The Analyze phase focuses on identifying and validating the root causes of the problem.

Answers

393. Answer: D

Explanation: The Sustain phase focuses on ensuring that the improvements are maintained over time and that environmental standards are upheld.

394. Answer: B

Explanation: Using fossil fuel heaters is not an example of a climate adaptation measure for houses; instead, renewable energy sources should be used.

395. Answer: C

Explanation: Motorola is recognized for pioneering Six Sigma in the 1980s, which helped in improving its manufacturing processes by reducing defects.

396. Answer: C

Explanation: Candidates must achieve a minimum score of 70% to attain the professional designation of IASSC Certified Green Belt (IASSC-CGB™).

397. Answer: CExplanation: Green Six Sigma focuses on reducing waste and increasing efficiency in supply chains.

398. Answer: A

Explanation: Walmart's supply chain expertise allowed for faster and more effective disaster relief during Hurricane Katrina compared to FEMA.

399. Answer: D

Answers

Explanation: Garvin's model includes dimensions like reliability, durability, and serviceability but does not specifically mention cost reduction as a quality dimension.

400. Answer: B

Explanation: Dow Chemical applied Six Sigma to improve its subsidiary FilmTec Corp, demonstrating the methodology's effectiveness in various industries.

401. Answer: C

Explanation: The "Electricity and other energy" sector is the largest contributor to greenhouse gas emissions. This sector includes the generation of electricity and heat by burning fossil fuels such as coal, oil, and natural gas. These activities release significant amounts of carbon dioxide (CO_2) and other greenhouse gases into the atmosphere. The high emissions from this sector are due to the extensive use of fossil fuels for power generation worldwide. Transitioning to renewable energy sources such as wind, solar, and hydropower is essential to reducing emissions from this sector and combating climate change.

402. Answer: C

Explanation: The final step in the process mapping procedure is to apply 'what if' analysis and simulation to achieve sustainable process improvement.

403. Answer: B

Explanation: TQM is defined as a management approach centered on quality, involving all members of the organization and aiming for long-term success through customer satisfaction.

Answers

404. **Answer: B Explanation**: The first step in creating a Gantt chart is to identify the key activities or tasks related to the project.

405. **Answer:** C

Explanation: The Monte Carlo Technique (MCT) is associated with the 'random walk' method, which involves random sampling to solve problems.

406. **Answer:** A

Explanation: SMED is a methodology developed by Shigeo Shingo to reduce changeover times in manufacturing processes.

407. **Answer:** C

Explanation: The Analyze phase aims to identify and validate the root causes of the problem.

408. **Answer:** C

Explanation: Smart glass windows are mentioned as a design feature to adapt houses to heatwaves.

409. **Answer:** D

Explanation: Professionals who successfully complete the certification process receive the designation of IASSC Certified Green Belt (IASSC-CGB™).

410. **Answer:** B

Answers

Explanation: DFSS (Design for Six Sigma) is suggested for use in accelerating R&D projects on alternative fuels.

411.   **Answer:** C

Explanation: The Osaka Blue Ocean Vision plan aims to achieve zero marine plastic litter by 2050.

412.   **Answer:** C

Explanation: The Balanced Scorecard is an enabler used in the Sustain phase to ensure a continuous improvement culture.

413.   **Answer:** A

Explanation: Flow process charts are symbolic representations of a physical process and its sequence of operations.

414.   **Answer:** B

Explanation: Juran emphasized the importance of continuous education and training to achieve quality within an organization.

415.   **Answer:** B

Explanation: The "Bring Back Mangroves Project" aims to support fisheries, provide food sources, and protect coastlines from flooding and erosion.

416.   **Answer:** D

Explanation: Control charts are used to monitor and sustain improvements in the Control phase.

Answers

417. Answer: B

Explanation: Brainstorming is a method used in the Improve phase to generate ideas and solutions to solve problems.

418. Answer: C

Explanation: Six Sigma Green Belts lead improvement projects and also serve as team members on more complex projects managed by Black Belts and Master Black Belts.

419. Answer: C

Explanation: Green Six Sigma uniquely integrates environmental sustainability into the traditional Six Sigma framework, focusing on both process improvement and ecological impact.

420. Answer: B

Explanation: Bangladesh developed a layered adaptation plan with an early warning system and concrete cyclone shelters, significantly reducing fatalities during cyclones.

421. Answer: C

Explanation: The measures for climate adaptation in retrofitting houses include reducing energy losses, replacing fossil fuel systems, using renewable energy, and applying a circular economy.

422. Answer: B

Explanation: SPC focuses on quality improvement through the use of statistical methods and tools to control process variability.

Answers

423.   Answer: B

Explanation: Quality 4.0 emphasizes the use of digital tools and real-time data to enhance quality and operational excellence in the context of the digital revolution.

424.   Answer: C

Explanation: The affinity diagram is used to categorize verbal information into an organized visual pattern and is often used during brainstorming sessions.

425.   Answer: C

Explanation: Mobile technology improves accessibility and communication within and across organizations.

426.   Answer: C

Explanation: The transportation industry is responsible for emitting 14% of greenhouse gases.

427.   Answer: C

Explanation: A Gantt chart is used in the Control phase to monitor project schedules and track progress.

428.   Answer: B

Explanation: The Process Capability Index (Cp) determines the reliability of process tolerances to meet customer specifications.

Answers

429. Answer: C

Explanation: EVM provides variance analysis and forecasting capabilities, helping to identify areas that need corrective action.

430. Answer: C

Explanation: A Green Belt Exam Voucher can be purchased for USD 350 through the provided link.

431. Answer: B

Explanation: Unilever's climate adaptation strategy includes investing €1 billion over 10 years in projects focused on landscape restoration, reforestation, carbon sequestration, wildlife protection, and water preservation.

432. Answer: C

Explanation: A study found that a hybrid fuel composed of 75% methanol and 25% hydrogen reduces carbon emissions by 65% compared to fossil fuels.

433. Answer: B

Explanation: Green Six Sigma extends DMAIC to DMAICS, adding a Sustain phase to ensure long-term success and environmental sustainability.

434. Answer: C

Explanation: Green supply chains focus on enhancing resilience to adapt to changing business conditions caused by climate change and other factors.

Answers

435. Answer: B

Explanation: SPC involves the analysis of 'special' and 'common' causes of variation to improve process stability and capability.

436. Answer: B

Explanation: The nominal group technique is designed to allow all team members to rank issues without pressure, giving quieter individuals the same influence as more dominant members.

437. Answer: C

Explanation: While quality assurance is important, it is not explicitly listed as a component of FIT methodology, which focuses on fitness for purpose, improvement and integration, and sustainability.

438. Answer: C

Explanation: Developing an in-house assessment process helps the company understand and meet its specific needs and criteria.

439. Answer: B

Explanation: The DMADV (define, measure, analyze, design, verify) program is a modified version of DMAIC used for more effective research in carbon capture technology.

440. Answer: B

Explanation: The COP26 climate conference, held in Glasgow, is referred to as the world's "last best chance" to reboot climate change targets and initiatives.

Answers

441. Answer: B

Explanation: Lean manufacturing focuses on just-in-time production, reducing inventory levels and ensuring materials arrive exactly when needed.

442. Answer: B

Explanation: The fundamental strategy of climate change preparation involves predicting longer-term causes and effects, preventing root causes such as greenhouse gas emissions, and preserving by adapting to changes despite preventative efforts.

443. Answer: D

Explanation: Scatter diagrams show potential correlations, not exact cause-and-effect relationships.

444. Answer: B

Explanation: Quality 4.0 relies on digital tools and systems rather than manual processes.

445. Answer: B

Explanation: Using self-adhesive foam strips for doors and windows is a common, inexpensive method to prevent draughts.

446. Answer: B

Explanation: A heat-only boiler (also known as a conventional boiler system) is suitable for a larger home with greater demand for hot water.

Answers

447.   Answer: B

Explanation: The activity on arrow method is most widely used for constructing an activity network diagram.

448.   Answer: C

Explanation: TPM focuses on improving manufacturing efficiency by eliminating losses and involving every employee in the maintenance process.

449.   Answer: C

Explanation: The initial assessment aims to understand the specific needs and align the program accordingly.

450.   Answer: B

Explanation: The first step in creating an affinity diagram is to clarify the problem or opportunity in a full sentence.

451.   Answer: C

Explanation: GE reported savings of $2 billion in 1999 due to its comprehensive Six Sigma program, illustrating its significant financial impact.

452.   Answer: B

Explanation: Brainstorming is a tool used in the Improve phase to develop ideas and solutions for the identified problem.

Answers

453.   Answer: D

Explanation: MSA assesses characteristics such as bias, linearity, stability, repeatability, and reproducibility, but not profitability.

454.   Answer: B

Explanation: Expanded polystyrene sheets are commonly recommended for insulating floorboards on the ground level of houses.

455.   Answer: C

Explanation: CEDAC (cause and effect diagram assisted by cards) is a variant of the cause and effect diagram.

456.   Answer: C

Explanation: The most genuine method of using renewable energy for a household is to consider domestic solar energy.

457.   Answer: D

Explanation: The 5S methodology includes Sort, Set in place, Shine, Standardize, and Sustain as part of its approach to workplace organization.

458.   Answer: B

Explanation: The main disadvantage of a GSHP is its high installation cost.

459.   Answer: B

Answers

Explanation: In the "Check" stage of the PDCA cycle, the results after implementation are compared with targets to assess if the expected performance improvement has been achieved.

460. **Answer:** A

Explanation: Green Belts support Black Belts and lead smaller-scale projects.

461. **Answer:** B

Explanation: RIBA (the Royal Institute of British Architects) has advocated for the preservation and re-purposing of buildings.

462. **Answer:** B

Explanation: Suppliers should be obliged to produce and supply spare parts for 10 years to facilitate repairs.

463. **Answer:** C

Explanation: The Project Charter outlines the project's scope, objectives, and benefits, providing a clear direction for the project team.

464. **Answer:** B**Explanation**: A circular economy aims at eliminating waste and the continual use of resources.

465. **Answer:** D

Explanation: Feigenbaum's cost of quality categories include prevention costs, appraisal costs, internal failure costs, and external failure costs, but not marketing costs.

Answers

466. Answer: B

Explanation: The pay-back period of a solar energy system is usually 8 years in the UK.

467. Answer: C

Explanation: The Five S's tool focuses on improving the housekeeping and organization of an operation, promoting efficiency and safety.

468. Answer: B

Explanation: The term "Six Sigma" refers to the statistical concept of standard deviation, with the goal of a process having Six Sigmas (standard deviations) between the mean and the nearest specification limit.

469. Answer: B

Explanation: DOE often uses orthogonal arrays to systematically study the effects of multiple factors on a process outcome.

470. Answer: C

Explanation: A flow diagram is used to identify potential failure points in a process during the Measure phase, helping to pinpoint where issues may arise.

471. Answer: C

Explanation: A histogram displays the frequency distribution of data.

Answers

472.   Answer: A

Explanation: Run charts are recommended for detecting visual trends and identifying areas for further analysis.

473.   Answer: C

Explanation: DMAIC Lite combines Define, Measure, Analyze, Improve, and Control into three stages.

474.   Answer: C

Explanation: ASHPs have the key advantage of lower energy bills compared to comparable gas heating systems.

475.   Answer: B

Explanation: GSHPs are known to be more energy-efficient and cheaper to run compared to ASHPs.

476.   Answer: A

Explanation: A radar chart is used in the Control phase to monitor the performance profile and track key metrics.

477.   Answer: B

Explanation: A milestone tracker diagram is used to show the projected milestone dates and the best-estimated date on the week of the progress review on a single chart, assessing the achievement or slippage of the progress.

478.   Answer: C

Answers

Explanation: As per the IASSC Recertification Policy implemented on March 1, 2017, certifications are classified as "Current" for three years.

479. **Answer:** B

Explanation: The Sustain phase in Green Six Sigma, often referred to as the Control phase in the broader Six Sigma methodology, focuses on maintaining the improvements achieved during the earlier phases of the project. This involves implementing the solutions and making sure they are consistently applied over time.

480. **Answer:** C

Explanation: An ASHP costs between $8,000 to $18,000 to install in the UK.

481. **Answer:** C

Explanation: A stand-alone electric water heater is recommended for small apartments.

482. **Answer:** A

Explanation: SPC stands for Statistical Process Control, a method for monitoring and controlling a process.

483. **Answer:** B

Explanation: Mineral wool, sold in rolls like a blanket, is the most common loft insulation material.

484. **Answer:** B

Answers

Explanation: The 5S methodology focuses on Sorting, Setting in order, Shining, Standardizing, and Sustaining to create an organized work environment.

485. **Answer:** B

Explanation: Installing a smart thermostat is suggested to help reduce energy consumption.

486. **Answer:** B

Explanation: TRIZ stands for Teoria Resheniya Izobreatatelskikh Zatatch, which translates to inventive or innovative problem-solving.

487. **Answer:** C

Explanation: The Sustain phase is the final step in the DMAICS cycle, ensuring that improvements are maintained and environmental standards are upheld.

488. **Answer:** A

Explanation: E-commerce involves transactions conducted electronically over the Internet.

489. **Answer:** B

Explanation: A histogram is used to show the frequency distribution of data, making it ideal for displaying defect occurrences.

490. **Answer:** B

Answers

Explanation: While Six Sigma Green Belts are involved in various roles related to process improvement and project management, a Marketing Director role is not typically associated with the technical and analytical focus of Six Sigma Green Belt certification.

491. **Answer:** C

Explanation: A Kaizen Event is a structured, rapid improvement process aimed at specific business aspects.

492. **Answer:** D

Explanation: Six Sigma is not one of the Five S's. The Five S's are Seiri, Seiton, Seiso, Seiketsu, and Shitsuke.

493. **Answer:** D

Explanation: Parasuraman et al. identified dimensions such as tangibles, reliability, and responsiveness but did not include innovation as a dimension of service quality.

494. **Answer:** C

Explanation: Cause and effect diagrams, also known as fishbone diagrams, are used to identify the root causes of problems.

495. **Answer:** D

Explanation: In poorly constructed homes, 40% of heat is lost through external walls.

496. **Answer:** D

Answers

Explanation: The Sustain phase in Green Six Sigma focuses on maintaining process improvements achieved during the project. Key deliverables include sustaining the implemented solution, promoting a culture of continuous improvement, and ensuring environmental benefits are maintained. Marketing strategies, however, are unrelated to these objectives, as they fall outside the scope of process improvement and operational efficiency central to Six Sigma methodologies.

497.   **Answer:** D

Explanation: Japan has the highest proportion of households fitted with air conditioning at 85%.

498.   **Answer:** C

Explanation: Black Belts lead project teams and are trained in Six Sigma methodologies.

499.   **Answer:** C

Explanation: The main areas of energy consumption in building operations are heating, cooling, lighting, and cooking.

500.   **Answer:** C

Explanation: Residential building operations account for 72% of the total emissions from the building sector.

About Our Products

Other products from VERSAtile Reads are:

 Elevate Your Leadership: The 10 Must-Have Skills

 Elevate Your Leadership: 8 Effective Communication Skills

 Elevate Your Leadership: 10 Leadership Styles for Every Situation

 300+ PMP Practice Questions Aligned with PMBOK 7, Agile Methods, and Key Process Groups – 2024

 Exam-Cram Essentials Last-Minute Guide to Ace the PMP Exam - Your Express Guide featuring PMBOK® Guide

 Career Mastery Blueprint - Strategies for Success in Work and Business

 Memory Magic: Unraveling the Secret of Mind Mastery

 The Success Equation Psychological Foundations For Accomplishment

 Fairy Dust Chronicles – The Short and Sweet of Wonder

 B2B Breakthrough – Proven Strategies from Real-World Case Studies

About Our Products

 CISSP Fast Track: Master CISSP Essentials for Exam Success

 CISA Fast Track: Master CISA Essentials for Exam Success

 CISM Fast Track: Master CISM Essentials for Exam Success

 CCSP Fast Track: Master CCSP Essentials for Exam Success

 Certified SCRUM Master Exam Cram Essentials

 AZ-900 Essentials: Fast Track to Exam Success

Copyright © 2024 VERSAtile Reads. All rights reserved.
This material is protected by copyright, any infringement will be dealt with legal and punitive action.

www.ingramcontent.com/pod-product-compliance
Lightning Source LLC
Chambersburg PA
CBHW062312220526
45479CB00004B/1146